No Small Potatoes: A Journey

More than Meets the Eye

Elizabeth Johnston

SOUND
PROOF
PRESS

© *No Small Potatoes: A Journey, Elizabeth Johnston, 2008*

All rights reserved. No part of this publication may be reproduced or transmitted in any form or by any means, electronic or mechanical, including photocopy, recording, or any information storage and retrieval system, without permission in writing from both the copyright owner and the publisher or a license from the Canadian Copyright Licensing Agency. For an Access Copyright license, visit www.accesscopyright.ca or telephone: 1-800-893-5777.

Published by Sound Proof Press
Cover Design: Robert Kertesz
Photographs: Elizabeth Johnston and IPBN
Printed and bound in the U.S.

Legal Deposit, 2008, Bibliothèque Nationale du Québec and National Library of Canada

Library and Archives Canada Cataloguing in Publication

Johnston, Elizabeth, 1963-
No small potatoes : a journey / Elizabeth Johnston.

Includes bibliographical references and index.
ISBN 978-0-9780743-0-2

1. Genetically modified foods--Popular works. 2. Potatoes--Biotechnology--Popular works. 3. Johnston, Elizabeth, 1963-.
I. Title.

TP248.65.F66J645 2008 363.19'2 C2008-905066-5

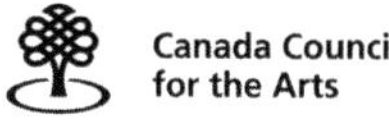

We acknowledge the support of the Canada Council for the Arts which last year invested $20.1 million in writing and publishing throughout Canada.

Sound Proof Press ♦ Montreal, Canada ♦ www.soundproofpress.com

For Barbara & Toni

Eternal vigilance is the price of liberty.

John Philpot Curran, 1750-1817

Dublin, Ireland

Table of Contents

Acknowledgements

Thank you to all my interviewees for generously taking the time to answer my questions. Your participation was invaluable. Thank you to those people who bought advance copies of *No Small Potatoes*. Your contribution afforded me some of that often rare commodity: time to write. In particular, thank you to those special people in my life who were my own private cheerleading team.

Introduction – Vegetable Love

Mine was a vegetable love, at first, because of my grandparents. Any affinity I have for the spud must be traced back to these hardworking Polish refugees from World War II who settled in Canada. They grew potatoes in their large hobby garden, and my memories of eating their potatoes are the strongest for me.

When it came to peeling potatoes, my grandmother, Barbara, was as skilled as a surgeon. Her paring knife took purchase just under the skin. She lifted the peel so quickly and thinly, it was as if it were only hovering above the white flesh. Deftly, she made short order of the spud's unassuming exterior. I would often watch her through the California window separating the kitchen from the living room, drooling at the thought of what would soon follow my grandmother's concentrated industry – perogies!

After she boiled the perogies, she'd fry them in a pan with butter and then plop them, sizzling, onto our plates. Sour cream, the fattiest kind you could get, was heaped onto the lightly browned crescents that moments before Barbara's nimble, mindless fingers had pressed closed, sealing in the secrets of the earth. The taste of melted butter, browned onions and the cool sour cream like a velvet shawl thrown over the perogies, in a dance of hot and cold, my mouth sang with the secrets of potatoes.

Recalling this makes me understand artist Joseph Beuys' comment that "even the act of peeling a potato can be a work of art," but my story might have ended there. That comforting memory may have been the one significant connection I had to the potato until the day I took a darkroom photography course. Testing out my new skills, I photographed all sorts of things including bricks, broken chairs and blenders. But when I put this lowly tuber between my camera lens and me, something mysterious happened. I remember it was a dark brown potato covered in dirt – not your typical *objet d'art* – and I wondered what on earth had possessed me to try to make art out of this lowly spud. As it sat there, silent as a monk, it occurred to me that there might be more to it than met the eye. What, I finally asked myself, were those secrets it had sung about way back in my childhood? What did it want to tell me now?

With those questions in mind, the potato led me on an eclectic journey that combined art, culture, self-determination, spirituality, organic farming, genetically modified organisms (GMOs), and our basic human right to safe, healthy food. Had anyone told me years ago that the potato was a fulcrum for all these issues as I scarfed down my granny's mashed potatoes, I would've said, "Pass the gravy."

Given my own attitude before embarking on *No Small Potatoes*, it's not surprising that people I encountered had wide-ranging reactions to the news about my spud odyssey. Both friends and strangers alike expressed curiosity, delight and sometimes skepticism. In fact, one friend even asked me, "What else do I *not* need to know about the potato?" I quickly realized that more than any other vegetable, the potato evokes strong reactions in people. As the head of communications for the International Potato Centre in Lima, Peru, put it, "No one gets worked up over lettuce like they do the potato."

We often take it for granted, though, because of its ubiquity. Since it is so common and abundant, it's practically invisible to most of us. Yet, it's at the centre of some significant changes that affect

- our access to food;
- how we relate to each other;
- how we relate to our environment;

- our health; and
- how we choose to live our lives.

These changes are brought about by the increasing corporate influence on agriculture. Global corporations are changing the face of the potato through mono-crops, factory farms, patents and genetically modified organisms (GMOs). These issues may seem far away from the concerns of most people today, especially in the Western World where the gap between rural and urban communities, and their respective lifestyles, continues to widen. But what is invisible to the naked eye can have the profoundest affect on our daily lives.

In *No Small Potatoes,* I explore these issues through research and interviews. In Peru, I visited the International Potato Centre (CIP) and the Potato Park. I also conducted interviews with Raymond Loo, organic farmer from Prince Edward Island, Canada; Arpad Pusztai, food scientist; Alejandro Argumedo, native Quechua Indian and agronomist from Peru. My research assistant, Fred Graham, conducted the interviews in Northern Ireland with farmer John Shepherd, agronomist Stephen Bell and others.

....

It is said that a nation's fate is determined by how it eats, and the potato is in a perfect position to demonstrate that.

Chapter One – Disappearing Acts

When I think of farms, I don't think of corporations; I think of hardworking individuals working the land. I think of the farms on the outskirts of the town my grandparents lived in, Wasaga Beach, Ontario. I think of green fields as far as the eye can see, the pungent smell of cows and the fan of geese and chickens as my grandfather and I approached the independent egg farmer's house. Increasingly, that idyllic childhood landscape is being transformed into factory farms where often only one variety of a crop is grown, and that mono-crop is usually destined for a manufacturer or distributor, not local markets.

Factory farms are mostly chemical farms where pesticides and herbicides are routinely used to protect crops from insects and diseases. Although using chemicals to kill weeds and insects has been a common practice since the mid-1900s, it's also been common to use natural seed, often saved from previous harvests. This seed-saving practice has gone on for as long as agriculture has existed. Many factory farms grow crops from natural seed. More and more though, factory farmers are contracted to grow genetically modified (GM) seed. These are seeds that have been altered at the gene level (to resist pesticides, for instance, amongst other things) and are protected by patents held by the corporation that developed the GM seed.

As factory farms get bigger, so does the use of GM seed. One of the biggest seed corporations worldwide is Monsanto, the American-owned company responsible also for the invention and distribution of the highly toxic chemical defoliant Agent Orange used in the Vietnam War. In 1988, Monsanto wasn't even on the Top Ten list of seed corporations, though it ranked seventh for global pesticide sales.[1] In 2004, it shot to the top, ranked as the number one seed company in the world, and its "biotech seeds and/or trait technology accounted for 88 percent of the total GM crop area worldwide. According to Monsanto, its biotech trait acreage covered 175.7 million acres in 2004 – roughly the size of Zambia."[2] More recent numbers given in the 2008 documentary, *The World According to Monsanto,* put Monsanto's genetically modified crops as covering 250 million acres worldwide. This coverage took only 10 years for Monsanto to attain. In the process, lives have been turned upside down, not to mention the agro-business of entire countries.

Pursuing a mandate to dominate the world seed market, biotech companies are endangering the global food supply as well as the environment.[3] Use of genetically modified seed and the practices of factory and mono-crop farming have profound effects on us all. So far, the effects have not been felt in the Western World as acutely as they have in the Third World, but that may be only a matter of time as the potato's story weaves in and out of this global David and Goliath drama. In short, the drive for supremacy in the GM seed market widens the gap between rural and urban communities, interferes with the transmission of knowledge, reduces diversity and affects quality of life. Listening to what the spud has to say about this crusade to seed the world with genetically modified organisms could mean the difference between subjugation to corporate objectives and individual independence.

---

[1] Vandana Shiva, *Monocultures of the Mind* (London: Zed Books, 2000) 128.

[2] "Global Seed Industry Concentration – 2005," *ETC Group Communiqué,* September/October 2005 Issue # 90, 6 Sep 2005 ‹http://www.mindfully.org/Farm/2005/Global-Seed-Industry6sep05.htm›

[3] *The World According to Monsanto,* dir. Marie-Monique Robin, Arte France, 2008.

Desire for Independence

The potato business, like every other business, exists within an established economic infrastructure, one that (with Free Trade Agreements, Patent Technology and the rise of the corporation) has industrialized farming practices to a large degree. No matter what the economic landscape looks like, however, the desire for independence remains constant. Two farmers, an ocean apart, demonstrate the universality of that urge: John Shepherd and Raymond Loo. John is a non-GMO, chemical potato farmer in Northern Ireland and Raymond is an organic potato farmer from Prince Edward Island in Canada. Though they differ in their growing approach, they both strive for independence and face the challenges that this choice carries.

Stephen Bell, an agronomist from Greenmount Agricultural College in Antrim, Northern Ireland, explained how the potato farming system works in his country.

"Pre-packers like Wilson's will have their dedicated growers. They've got about 20 growers in Northern Ireland supplying all of the potatoes they require, and they're not willing really to let too many other people join that club, because they want to keep the quality high and they want to be in control of it."

In Prince Edward Island, many of the farmers produce for the Cavendish frozen food factory. Independent farmers like organic potato grower Raymond Loo are dwarfed by the huge factory farms that surround his small farm.

Niche Markets

As an independent farmer by choice, Irishman John Shepherd is outside the vertical integration that guarantees a market for his potato yield. His market is on a smaller scale; one that he and his wife service. It's a choice that guarantees autonomy but also comes with the challenge of maintaining a personal market niche. He sells locally, as do many other small potato growers.

"They'll sell them at [people's] houses or at smaller, local supermarkets," said Bell, "People will come to them [because the small farmers] have a reputation for producing good potatoes."

Raymond Loo, PEI organic potato grower, cultivates that reputation by selling at local farmers' markets where he has regular kiosks. There he can talk one-on-one with his loyal, repeat customer base as well as introduce them to specialty, hard-to-find and heritage varieties of potatoes that big supermarkets won't carry.

Both John and Raymond have used their entrepreneurial freedom to carve out niche markets for themselves despite being surrounded by factory farms. Even though large corporate interests overshadow them, there are still strong farmers' associations and citizen rights' groups that make their voices heard. Democratic rights like these are a given in First World countries, but not so in some Third World countries. While John Shepherd and Raymond Loo have the option of carving out niche markets, some farmers, like those in Paraguay, are not so lucky.

Paraguay's Story

Paraguay is a largely agricultural country where most farmers used to be small, subsistence farmers, able to feed themselves and their families until factory farms of GM soybean took over the countryside. Although Paraguay was officially opposed to GM plants, Monsanto seed somehow infiltrated the country and bred through cross-pollination with natural soybean seed. In addition, the Monsanto chemical herbicide Roundup Ready that was sprayed over the crop killed the surrounding plants whether they were weeds or not. "Roundup is a non-selective herbicide that does not distinguish between weeds and desirable vegetation, and thus kills all plants, which is in no way economical."[4] The only plant not affected by the herbicide is the genetically modified soybean.

In Paraguay today, "70 percent of the land is owned by only 2 percent of the population. With GMOs, the concentration is increasing. Three quarters of the soybean producers are foreigners staking claims for this new green gold," said Marie-Monique Robin, director of the documentary *The World According to Monsanto.*

The small farmers who remain must pay Monsanto the royalties it demands for using its patented seeds, whether they bought the seed or it was blown by the wind onto their property. Once the GM soybean mingles with natural soybean, there is no going back to "pure" soybeans. If farmers are unable to pay the royalties, they can lose their farm. The resulting displacement has torn communities and families apart. Many have had to go to the cities, but conditions are even worse there with no work, or home. One displaced farmer put it this way: "In my case, my family lives in the city, but I don't want to go there. In the city, you have to buy everything, even food. Here whatever we grow is ours. We can eat whatever we want. But in the city, you can't. If you don't have money, you have to search for food in the garbage cans."

The destruction of families and community infrastructures are a by-product of corporate expansion at all costs. Marilyn Waring, author of books that investigate the inequities of the current biased economic system, tells the story of how a rural Philippine family were affected by the construction of a new factory. Before the factory was built, families were poor by Western standards but had food, clothing and community. The promise of well-paying jobs lured people from their village, however, and when the factory closed only a few years later, there was no longer a

[4] Vandana Shiva, *Stolen Harvest* (Cambridge: South End Press, 2000) 98.

village to come back to. Everything these villagers had was lost and they were now lowered to finding food in garbage cans, just like those in Paraguay who were displaced by Monsanto's corporate takeover of agriculture though GM soybean.

"What we have here," said Jorges Galhario, leader of a small farmers' organization fighting against the GM soybean invasion of Paraguay, "is a mono-crop that destroys everything in its path. Before, here there were fields containing everything that a family needed to live: plants, trees, manioc, and corn." To Jorges, it's obvious that small farmers cannot exist within an economic system that allows corporations such as Monsanto to flourish like a herbicide-resistant weed. "They are two incompatible models that can't co-exist. It's a silent war that eliminates communities and families of small farmers," said Jorges.

A recent census from Paraguay supports the anecdotal. "Each year, 100,000 people leave rural areas to live in urban slums. An estimated 70 percent flee Monsanto's genetically modified soybeans, which are destined to feed Europe's chickens, cows and pigs."[5]

The enormity of the devastation wrought by Monsanto and GM technology boggles the mind. Admittedly, Paraguay is an extreme case, and when faced with an extreme, the usual reaction is disbelief. *Surely this wouldn't happen here?!* But maybe it is, only in a quieter way and at a much slower rate.

John Shepherd, Northern Ireland farmer, has noticed the change his farming community has undergone since he was a child. "When I was small," John remembered, "there were always wee tractors floating about, people floating about … [Now,] the tourists wouldn't have anybody to talk to."

---

[5] *The World According to Monsanto.*

In Prince Edward Island, Raymond Loo's small organic farm is dwarfed by huge factory farms. In fact, I noticed the strange silence of the countryside on my way to the PEI potato museum. The drive took me through the Cavendish Farms sector of the island. There were miles and miles of green fields unrelieved by farmhouses or villages, only the occasional factory where farmers would deliver their produce. It didn't feel like a rural community, more like a huge sprawling factory. The words of Jorges the farmers' activist echoed in my mind. "[Monsanto] is a true multi-national company. It's everywhere in the world. Its objective is to control all of the world's food production through farmer-less farming."

As of my visit to PEI in 2006, there were no GM crops being grown, according to Ivan Noonan of the PEI Potato Board, but as I drove through that "green desert" breathing in toxic chemicals for the whole way, an eerie feeling came over me. The gap between urban and rural communities was profoundly tangible. It tasted acrid on the tongue and my eyes stung with its monotony.

Chapter Two – Well, What do You Know?

Variety is said to be the spice of life, but not only does mono-crop agriculture alter the way we relate to the land, "it [also] destroys the biodiversity of the countryside," according to Jorges Galhario.

But what does biodiversity really mean? That depends on where you live. For consumers like me who live in the city, the threat to biodiversity means less choice and taste, and for Westerners that's very significant. It almost seems to be one of our democratic rights – choice and easy access to that choice. Yet, in the Canadian groceries stores I frequent, there are only a handful of potato varieties that are regularly available. We have reds, whites, yellow fleshed, and baking potatoes, none of which are identified by the variety name. On the other hand, in Peru, there are over 5,000 naturally occurring varieties of potatoes. That means you could eat a different type of potato every single day for over 14 years. Since Peru is the origin of the potato and possesses the largest genetic potato pool in the world, you might think that in a big city like Lima, your Peruvian potato sack would spill over with the vast array of varieties. Not so. In the downtown grocery stores I went into, there were only four varieties to choose from. According to Raymond Loo, PEI potato farmer, the dearth of choice is because those varieties are "the ones easiest to grow" on factory farms. Ironically, even in Peru, mono-crop agriculture limits consumer choice, and where the potato is concerned, that is an enormous curtailment of humanity's access to nature's abundance for food, remedies, and problem solving.

Different Varieties

In my local supermarket, we do sometimes get new varieties. More often, though, I have to go to farmers' markets to get something different like organic or blue potatoes, a local variety that has been cropping up lately. But even then, the variety name is still not listed. While it's great to be getting access to new potato varieties, they don't always come with instructions, and people don't always want to try new things unless they're sure they know what to do with them.

This lack of knowledge and/or our suspicion is something that small farmers deal with on a daily basis. Partly it's do with the labeling system, and partly it's to do with people thinking they know all there is to know about potatoes. A spud is a spud, afterall, even in Ireland. When people in a major Belfast grocery store were asked why they were buying those particular potatoes, the reply was, "Well, they're potatoes." However, not all potatoes are equal.

"The problem is," said PEI farmer Raymond, "you can walk into the store and buy round whites one day, and they're [the variety called] Superior. And the next one they pick up might be Atlantics, [or] ... Kennebecs, or something. But they all cook differently. Well, if you're not told that they all cook differently, then you have three potatoes that are left in a bag from the week before, and you just throw them in a pot with the others, and some of them fall apart and the others are still so hard."

"You know, I've had people come to the market, [saying], 'Well, I just don't think I'm a good cook. I just can't seem to get potatoes to turn out.' Well, that's the [fault of the] people who are putting them in the bag, not your fault," Raymond would say to them. "You cook them this way and they'll turn out. And people come back next week [with a] big smile. So the potato industry has done a terrible job of trying to tell people how to cook them."

Part of the small farmers' job, as Raymond Loo sees it, is to fill in the knowledge gaps that have resulted from the industrialization of farming. Mass mono-crop agriculture has affected the flow of information between rural and urban dwellers but also intra-rural, too. Proud of his potatoes, Raymond gave some to a long-time friend and neighbour only to find out later that his friend hadn't even tried them.

"Island Sunshine is, of course, really, really, really yellow," said Loo. "I gave them to one woman. It was so funny. She was a friend of mine, but I didn't hear back from her for a long time."

Eventually, Loo approached his friend and asked her how she liked them. She told him that she had to throw them out because "they were all yellow inside."

"She threw every one of them away!" said Raymond Loo. "I said, 'Of course; they're supposed to be.' She'd never seen a yellow potato before. She grew up here, and thought it was some kind of terrible disease and didn't want tell me."

It's a common problem for Raymond – people not wanting to try something they haven't already been introduced to. "But you know," said Raymond, "if their grandfather didn't eat them, by golly, they're not gonna try 'em, either."

Agronomist Stephen Bell has noticed the same thing in Northern Ireland. "People would've been brought up in the country on these potatoes, and their attitudes won't change. Their parents cooked Kerrs Pinks; they'll continue to eat Kerrs Pinks."

"It's kind of a frustration trying to get new stuff onto the market," said Raymond. For him, the unwillingness on the part of the consumer to try new things is just as challenging as overcoming their misconceptions. To try to counter some of that resistance, Raymond plans to put labels on his potato bags with the variety name and how to cook them.

"The next thing we're going to do is put up a list of varieties and stick it on the bag," said Raymond, "and it'll say this is this variety and this is how you should cook it." For example, the Loos grow a variety called Splash. Its skin is purple and white and its flesh yellow. Raymond said it's wonderful sliced and thrown in a pan with some olive oil and garlic. "If I want people to be really successful at their meal, that's what I've got to do. Give them the tools so they can be."

Being a small independent farmer means you're in charge of a lot more than just growing spuds. Marketing and promotion go hand in hand for the independent farmer. However, Switzerland is already putting more information on their spud bags, and in the UK, the Marks and Spencers stores even put a picture of the farmer on its gourmet potatoes.

Green Potatoes

Another necessary tool for potato-eating success is also just knowing basic things like the fact that green potatoes aren't good for eating.

"If you go into the supermarkets," said Irish farmer John Shepherd, "everything's sitting under bright lights. The soil's washed off of them ... and in two or three days they are green."

Though it may seem common knowledge that green potatoes aren't to be eaten (as they can cause stomach upset), not everyone seems aware of that as this experience recounted by agronomist Stephen Bell attests to.

"I had a group of people in the fields one day," recalls Bell, "and this 40ish-year-old woman, she worked for the civil service, was from the town, and she obviously didn't know anything about [potatoes]. She said 'I've bought potatoes and [when] they're green, I've left them out to ripen.' I tried to tell them; they're green from being in contact with light, and they're not going to get any better. So throw them out. There's so much people don't know."

The fact that so many people live in cities, divorced from the country where their food gets produced, means that they don't have as much knowledge about how things are produced, and why they're produced that way.

"That's one of the issues now in our society. We've been completely disconnected from our food source," said Toronto food bank executive, Nick Saul. "Kids in particular these days are at a complete loss [here]. They think that tomatoes grow on supermarket shelves."[6]

Possessing a minimum of food knowledge is important because it leads to making better choices. Our society may be about democracy and free choice, but the quality of our choices is dependent on the quality and availability of knowledge. At first glance, it may seem immaterial where we get our information and who is giving it. Yet, it does make a difference because the information we are getting from businesses, and increasingly governments, is informed by corporate interests, not health ones.

Recently there have been television commercials from Ontario Foodland informing the public on how to keep mushrooms fresh. This information would be common knowledge if a connection between the city and the country still existed. But without it, 'experts' in the form of marketing boards and businesses educate in order to sell more. Instead of the emphasis being on how best to feed ourselves, the information increasingly comes from how best to sell us products.

Eyes

Another misconception that stems from lack of exposure to those who grow our food is the belief that spuds with a lot

---

[6] Nick Saul, "The Green Barn," video ‹www.thestop.org›.

of eyes are not good for eating. When asking shoppers in Antrim Town, not far from Belfast, about an old variety he remembers his mother cooking, Arran Victories (also known as Irish Blues), the people my researcher Fred spoke to said, "They don't look too good. There's lots of eyes."

In North America, we may not give potatoes with multiple or pronounced eyes a second look, but in Peru, many of the potato varieties that I saw in the markets had a lot of eyes. For the native Peruvians, the Quechua, the eyes are thought of as teeth, and they have a special significance. The idea is that teeth are needed to carve a place in the earth, not eyes, and so the more "teeth" a potato had, the better chance it would take root. As well, the number of teeth is thought to correspond to the number of potential potato offspring.[7] In other words, the more teeth a potato had, the more abundant it would be.

Lacking such history and connection to the spud, most consumers in developed nations reject potatoes that give an eyeful. Because of consumer reaction like that, small grocery owner, Desi Boyle of Antrim, carries potatoes that are easy to cook, such as Maris Pipers and Navans. "The blues," Fred recalled of his discussion with Desi, "didn't sell a lot because they had too many eyes; they were too difficult to take care of. Because of the number of eyes, it's the kind of potato you need to cook in its jacket, and then peel afterwards."

More and more today, convenience drives consumer choice, and that is a factor in what Raymond McCafferty, a Co. Antrim potato distributor to restaurants, carries. "Well, there's too many eyes in them for the catering," said Raymond. "I wouldn't be handling them, but they're excellent potatoes. They're really dry. Kerrs Pinks and Arran Banners; they're Blues."

---

[7] Radcliffe N. Salaman, *The History and Social Influence of the Potato* (Cambridge: University Press, 1949) 24-5.

John Shepherd, too, liked the Blues, but recognized their limited appeal in the marketplace. "Yes, well, the eyes are the thing," said John. "That's a problem. The best spuds are your Queens, your Kerrs Pinks and your Blues for eating [all of which have a lot of eyes. So,] the best looking potato is generally the worst eating potato."

A Matter of Taste

When the biodiversity of a crop is affected, so is the range of tastes. If you reduce the available choice of potatoes, it makes logical sense that taste will also be affected, if only because there are fewer varieties to excite your mouth with. But the way a potato is grown also affects the quality of its taste. Both Raymond Loo and John Shepherd recognize the demands of the pre-packers and supermarkets to produce potatoes of long shelf life, but in doing so, according to Raymond and John, taste is sacrificed.

Before the Loo family went the organic, independent route, Raymond's father met with Cavendish Farms representatives in the early 1990s to talk about being agents for the Loo variety, Island Sunshine. Loo Senior was very proud of his new variety for many reasons, not the least of which was its taste.

"The fellow told him, that doesn't matter," recalled Raymond. "People just put salt and ketchup on them. But Dad felt it *should* taste good. Potatoes should taste good. That should be an attribute of its worth. So, once in a while these lights were coming on for us, and we were saying there's something wrong with this whole potato industry that taste doesn't matter. Where we going with that? And I think from that point on, it became very important for all our potato varieties to taste good. First and foremost, they have to taste good."

For some people, the taste of potatoes isn't as discernible as it is for others – especially if you're just dousing them with ketchup. But many people I've talked to are adamant that not only does taste matter but that it has definitely changed over the years. A woman I know, who is originally from the Dublin area, remembers her father talking about potatoes in his retirement. He was a hobby gardener all his life and grew potatoes in his backyard until just before he died. He didn't know what the difference was, but he was sure that the potatoes he grew in his retirement didn't have the same taste as the ones he grew years before.

Fred Graham, my research assistant, also notices a distinct difference in taste amongst potatoes, in general, but also between what he used to eat as a child and what's available today. Being aware of this taste difference, it's one of the things Fred talked about with people in Northern Ireland.

"Have you personally noticed a difference in taste in potatoes?" asked Fred of Raymond McCafferty, the distributor who sells prepackaged potatoes and gourmet potatoes to restaurants,

"Oh, the County Down," Raymond immediately replied. "When I was young, say 13, 14, I remember my Granny McCafferty buying County Down potatoes, and they were absolutely beautiful. I don't know what has actually changed, but I know they're not the same flavour now. I would tend to persevere with the old Irish, the old season variety, they would be better to taste.

"The main Irish crop of potatoes," continued Raymond McCafferty, "what they call 'earlies,' would be County Down potatoes, golden County Down. They used to come out in the month of June. They used to be a very dry, floury potato. But in latter years, in my opinion and the general consensus is that, years and years ago, the County Down potato was a potato you looked forward to eating. But I think they've become a forced crop. They forced the crop to come up too early and so the potato suffered. It's a wet potato now."

To force a crop, potatoes are watered more often and given more fertilizer resulting in a wetter texture when cooked, but taste suffers. But, when you're concerned about volume like Cavendish farms in Prince Edward Island or Wilson's Country in Northern Ireland, having a steady supply of attractive spuds is more important than how good they taste.

NI agronomist Stephen Bell talked about the gap between what supermarkets want and what consumers prefer. "The major pre-packer in Northern Ireland, that would be Wilson's Country Potatoes," he said. "They're supplying Tesco's, and the main variety they'd sell is Cultra. It looks well when it's grown, a new variety they encourage growth of, but whenever you go to actually eat them, a lot of NI consumers, the older generation, are not that fond of them [because of the bland taste]. But they look good, and that's what the supermarkets are after.

John Shepherd noticed the taste difference in turnips, too. "We used to grow a turnip called Swedes, lovely nice moist-tasting turnip. You can't get the seed now, because the supermarket doesn't want it. There was too much sap in it. So, whenever you put it on the shelf, it dried out, and then they have come out with [these tasteless ones]. I don't know where they have developed these seeds from, and I don't know where they've brought these other turnips from but they would be domineering now; they are more profile shaped. They're hard as stone. Sheep won't even eat them. But they sell them because they have a long shelf life," John said.

Selective breeding to produce a potato that stays hard longer might please the supermarkets and pre-packers, but it doesn't always please the palate. In the end, it seems that the people who really benefit from a vegetable with a long shelf life are the stores and the pre-packers. They can store things longer and present them in a clean, fashionable light. Taste is a secondary concern. However, I think from a consumer standpoint, taste and nutrition are among the primary considerations.

Ideally, a potato that tasted as good as it had in the past and had a long shelf life would suit both the eater and the corporations. "That would be the ideal world," said Bell, "but [they] haven't as yet found that variety." However, with the narrowing of the gene pool and the reduction in biodiversity, it's not likely such a spud will be found.

Le Sexy Spud

A potato variety that I used to favour was Yukon Gold because of its slightly nutty flavour, but I don't eat them any more because it tastes bland to me now and too wet when cooked.

"Yukon Golds are a nice potato," said Raymond, "but again, I think a lot of the flavour comes down to actually growing them a little bit slower. See, our potatoes come up much slower. We don't have as much nitrogen in there. The potato has to actually spread a bigger root system to try and find its own nitrogen. It's something like free-range chicken, people will come in and say, 'That tastes like the chicken we had years ago.' Well, it's because the chicken is running around outside and they're getting all they need, and you're getting all the flavour there. Same thing with potatoes.

"Partly it's variety, but partly it's production," Raymond emphasized. "How do we actually make this taste better so that you come in and say, *Wow, these are good!*"

Hearing Raymond talk about how he grows his potatoes, paying them special attention, and giving them time to develop slowly, made me think of how much care goes into winemaking or cheese. In the same way, Raymond shows that there's an art to the growing of potatoes as well. Not only could it be a comfort food that filled up the belly, it could excite the taste buds in unexpected ways. Like a fine wine, no potato before its time ensures a sensual taste experience!

"I can see there'll always be a market for commercial, big production," said Raymond, "for French fries and probably the super stores that are selling a big ten-pound bag for a dollar. But there's also a market for people who are selling the ten-pound bag for 10 dollars. I mean, let's face it. That's only four large cups of coffee, at Starbucks. So, if you could make this food a little bit sexy, if you will, somehow make people think, *Boy, I want that.*"

Concentrating on growing potatoes that tempt the taste buds is a good way to make people take another look at the humble spud, and the idea of making the potato sexy has captured my imagination ever since. I realized, though, that even before Raymond suggested that, I was already open to new potato experiences. While in PEI, I made a point of going to a big hotel in downtown Charlottetown specifically to try their "Potato Martini" a delicious mashed potato appetizer served in a martini glass. Somehow, dressing the humble spud in the sleek lines associated with the martini did take it out of the ordinary. The smooth whipped potato enveloped tangy bits of scallion and freshly cooked bacon bits, the tastes mingled in my mouth in a sensuous tango. Given that really good tasting potatoes take a long time to grow, it seemed only right to savour this whimsical concoction. Ah, Le Sexy Spud!

Since taste is a gateway to the sensual, coming to think of the potato as sexy wasn't so difficult for me, but it's definitely second nature to the Quechua, natives of Peru who have cultivated the spud for thousands of years. They associate the potato with sexuality, fertility and human life, itself.

The Quechua believed that the potato crop's success was dependent upon the sexual procreation of the plants, and they performed blood rituals to ensure a bountiful harvest. The blood was poured onto the land to stimulate the Potato God towards successful fertilization and reproduction. For the Quechua, potatoes, as they believe do all plants, are animated by a spirit, and it's to this spirit that their rituals are directed. The fact that many Peruvian potatoes are blue and red inside with veins of colour running through them only reinforced the comparison between it and the human body.

Another aspect of the sexuality associated with spuds can be seen in the ancient Quechua water jugs made of pottery. The Larco museum in Lima, Peru, has an impressive collection. A large number of them are in the shape of potato tubers and often incorporate humans in sexual positions or hybrids of potato and human figures. One jug had the figure of a man's head with a bulbous, potato-like body. Out of his stomach and sides protruded other heads. Another jug, dated approximately 200 B.C., depicted two potato-looking people in the act of intercourse. These vegetable/human couplings reflected the Quechua belief in the inextricable bonding between humans and the Potato God who provided their sustenance. Teeth were often highlighted on these jugs, too, symbolically representing the Quechua potato eye/teeth belief. Many of the faces bore cleft lips and this was explained in Salaman's seminal book on the potato, as being connected to the ancient Peruvians' belief that teeth were what ensured good crops.[8] The more of them that could be seen the better, hence the excising of the lips on their pottery when depicting humans. It's also believed that people who were born with cleft lips were considered to be good omens for the success of the potato crop in the village.

---

[8] Salaman, 24-25.

Today in Peru, potatoes still hold an important place in spiritual and religious rituals. While in Puño, I visited the Uros, a native group that live on their floating reed islands in Lake Titicaca. There I saw an altar that had an egg, some seeds and tiny potatoes on it. In a Catholic church located in downtown Puño, I saw a woman place a silver platter of potatoes on a table below the front altar. The priest came down and blessed the woman and her offering. Then the altar boy took the spuds into the sacristy.

Part of the Family

So important to the Quechua culture are potatoes that they are given a respect and stature almost beyond First World imagining. As Quechua-Canadian Alejandro Argumedo put it, "Potatoes are like living beings. People treat them like that. They are members of the family for farmers."[9] Not only are they members of the family, they're considered equal partners in the ongoing dance of life where variety is the secret to solving some problems – agricultural or individual. It's a reciprocal exchange of care and information.

For the Quechua, communication works both ways, and that worldview is not too far away from the intimate relationship experienced by the Irish during the time of the *clachans*. The *clachan* was a land-sharing system that ensured equality for all in the community, according to historian Kevin Whelan.[10] This system flourished in pre-Britain Ireland where life was organized by the needs of the community. Everything was communal; ownership wasn't defined like it is today. Rather the land was owned by everyone, and

---

[9] Joanne Silberner, "In Highland Peru, a Culture Confronts Blight," 3 Mar. 2008 ‹www.npr.org/templates/story/story.php?storyId=87811933&ps=bb1›.

[10] Kevin Whelan, "Writing Northern Ireland: History, Memory and Politics in Heaney, Friel and Deane," Inaugural Colloquium in Canadian Irish Studies, Concordia University, Montreal, 22 Oct. 2007.

everyone shared equally in its riches and its scarcities. In this system, "land connects to the inner contours of the imagination," said Whelan. There was a "bottom-up evolution of culture – the language and the landscape" worked together to form the culture. The reciprocal nature of people working with the land, instead of merely using it, allowed for accumulated knowledge and the ability to solve problems both for plants and people. "Tradition [in the *clachan* system] was not stultifying. It preserved successful innovation and successful solutions to persistent problems." Herbal remedies and natural insect control are just two examples found in other cultures.

"In Indian agriculture, women use up to 150 different species of plants (which the biotech industry would call weeds) as medicine, food, or fodder. For the poorest, this biodiversity is the most important resource for survival."[11] In Peru there is a potato blossom used to make tea for urinary tract ailments, and the blue potatoes are known for their high antioxidant levels. In fact, these potatoes have more antioxidant in them than blueberries. Another potato, packed full of carotene, is good for preventing blindness. Some potatoes are also good as catch crops, which are plants that attract insects to them, thereby drawing them away from the desired crop. It's a natural way of controlling pests without the use of pesticides.

Raymond Loo came by this knowledge by accident and used that information to protect his preferred variety of potatoes. He found a bag of potato seed in the shed, which his Dad had saved, and then planted them, just to see what would spring up. He soon realized that the bugs were very attracted to those particular ones, and so he planted two rows as catch crops.

"The insects love them, so it's a very good way of organically farming," said Raymond.

---

[11] Shiva, *Stolen Harvest*, 104.

Plants that draw bugs away from other plants are also known as beneficials, and concentrated knowledge about these types of plants is contained almost exclusively in ancient communities such as the Mayans of Guatemala and the Native Quechua of Peru.

It was great luck that Raymond happened upon the discovery of a beneficial for his potato crops, but he recognizes that his farming business would profit from more of the same kind of knowledge, which the Quechua have been collecting for thousands of years.

"I bet the old people in that Potato Park have a lot of answers for that area. Because when you talk about preserving beneficial habitats, that's the kind of knowledge that I don't have," said Raymond.

Collective Knowledge

Though there is more information available today than there used to be for independent farmers, it's not always easy to find, and the knowledge is not as deep as that held in Native communities like the Mayans or the Quechua. As well, the centuries'-old practice of using beneficials relies on the availability of a wide variety of plants. But when mono-crop farming takes over an area, the level of knowledge farmers have at their disposal decreases or disappears all together. Bigger mono-crop farms mean fewer farmers and less biodiversity. Fewer farmers mean fewer fellow experts to exchange knowledge with, and bigger farms mean less time a farmer has to work with the land organically and safely.

"A farmer working a mono-crop doesn't have the time to really know his crops," said Raymond Loo. "What happens is you get a "dumbing-down" of farmers, if you will, because the farms get too big so you can't physically walk everywhere. You can't physically plant one plant beside another plant because you don't have time anymore."

As well, factory farmers must often follow growing instructions they receive from the pre-packer's in-house agronomist to the letter. If they don't, they run the risk of not getting the best price for their products, since the pre-packers want a consistent size and shape all the time. This leaves little room for individual experimentation, innovation, or getting to know what certain areas of the natural fields might be capable of.

"Fields are all different, and I have some fields that I wouldn't have in grass," said John Shepherd. "We'd have a wide variety in our soil type here. Like at home our auld wettest heaviest field is side by side with our lightest sandiest field, believe it or not." John pointed to a part of his land. "That hill I would very seldom put in potatoes because it just doesn't do well. And you'll find good land and different land, will suit different varieties."

John's farm has been in the family for over 300 years, going back to just after the Famines. But no potatoes were grown on the family farm until John inherited it. So, everything he learned, he learned the hard way. Growing potatoes, watching how they fared in this parcel of land or that parcel, paying attention, and getting personally involved with his work resulted in an intimate knowledge about the potential between land and farmer.

Because John is a small farmer, he's able to make decisions about his crops based on his comprehensively intimate knowledge of his land. His independence from factory farm instructions allow him the freedom necessary to problem solve in a way that meets the very specific needs of his land and himself – not those of a pre-packer interested in cookie-cutter products instead of natural produce. This same freedom is what allowed the Loo family of PEI to create a successful solution to the persistent problem of potato blight. They were able to freely use their knowledge of how the plant works and to take their time experimenting with many varieties so that they could be better farmers, not businesspeople getting rich off of other farmers by charging them user and royalty fees in perpetuity. Genetic modification stops such knowledge creation in its tracks.

Saving the World through Science

But if science can help along this creativity that small farmers come by naturally but slowly, wouldn't that be better for them? And wouldn't we have more food to feed the masses?

Helping people live better lives is one of the most powerful justifications for the propagation of genetically modified plants. It's common to hear biotech companies say that their products will benefit humanity and even solve world hunger. Since the 1950s when a new term was coined, the Green Revolution, the role of science in agriculture has increased. The so-called Green Revolution is a movement that started just after World War II with the objective of helping feed the expected population explosion of the near future. Now, however, it serves biotech companies more than it does humanity.

Companies that advocate the use of chemicals and genetically modified organisms (GMOs), do so because they say such technology means

- better yields;
- less pesticides;
- could lead to the elimination of blight; and
- that GM seed is the same as natural seed (substantially equivalent) and therefore safe to eat.

The fact of the matter is that many farmers are finding out that this is not the case.

Better Yields, Less Pesticides

In India, for instance, GM cotton has become more difficult to grow as the pesticides used on it (supplied by the GM seed company) are no longer effective on the weeds. This led to farmers having to buy more of the expensive pesticide, also available only through the seed company. In Paraguay, as seen in the documentary, *The World According to Monsanto,* small farmers who are resisting the spread of soybean factory farms are still being affected by the chemicals sprayed on the soy fields. Everything except the soy is killed by the chemicals because the soy has been genetically modified to resist the effects of the chemical. But people who live near the fields have become mysteriously ill. Two little boys who have to walk through a GM soy field to bring their mother's organic corn to town have developed rashes on their legs and lost their appetite.

Closer to home in Canada, Prince Edward Island has a higher than the national average of rare and aggressive cancers, according to Sharon Labchuk, PEI Green Party leader. Labchuk noted that the two most-used pesticides on PEI, chlorothalonil and mancozeb, are classified as causing cancer and neurological problems. Both are fungicides used on potato fields, sprayed as often as every four days, and represent more than one-half, by weight, of agricultural pesticides used on the Island."[12]

Though the provincial department of health says there's nothing unusual with the rate of cancers in PEI, a local doctor has a different opinion.

"Dr. Ron Matsusaki, emergency room physician at Western Hospital in Alberton, says that in all the years he's worked as a doctor both in Canada and the US, he hasn't seen cancer rates that come even remotely close to what he's seeing in the West Prince area of PEI. He says he has no doubt that these cancers are caused by 'an insane

[12] Press release, 17 July 2007, ‹www.greenparty.pe.ca/node/151›.

amount' of chemical pesticides. Every second household in Mimnegash, a fishing village in West Prince surrounded by potato fields, has been afflicted with cancer, according to Matsusaki."[13]

Raymond Loo has also been affected by the promises of "better farming through technology," the promise of the Green Revolution. His father, Gerrit, developed leukemia and, according to Raymond, the doctor said it was likely he got it by using the pesticides he worked with before the farm went organic.

"It was a little bit difficult for Dad to make this transition to organic because he was a big believer in the Green Revolution. Big believer in cutting down all the hedges and building bigger ... and by golly this here [method] would feed the world. 'Pesticides and plastics were good things.'

"We had these huge arguments," recalled Raymond. "I remember he would just stick his arm in the [pesticide] sprayer and mix everything up."

The rationale Raymond's father gave was that the pesticides he was using were safe because of the benign symbol on the container. "Look it's got a green cross on it," Gerrit would say. "That's not poison; it just kills plants."

Packaging chemicals with images that make them appear harmless is not uncommon. In India, one line of GM seed put out by Monsanto had a Hindu god on it, giving the impression, then, that these seeds were not only safe but blessed with superior abilities to produce high yields as well as resist certain insects, remarked activist Vandana Shiva.

Despite the increased reliance on science to produce genetically modified plants that are supposedly designed to deliver better yields and decrease pesticide use, the promises have not materialized. The situation of farmers

---

[13] Joan Delaney, "Potato Farms a Hotbed of Cancer?" *Epoch Times*, 23 Mar. 2006 ‹en.epochtimes.com/news/6-3-23/39627.html›.

in India using Monsanto's Bollgard cotton, a plant genetically modified to produce an insecticide that kills the cotton parasite, Boll worm, is a case in point.

India is the third largest cotton producer in the world. In 1999, Monsanto acquired its largest seed company, and two years later India approved the sale of its genetically modified cotton. A television advertisement about Bollgard that ran in India states, "Less herbicide, higher profits."

Noted physicist turned farmers' activist Vandana Shiva (and president of the Navdanya Foundation) has spent decades fighting first the introduction of industrialized farming into India and now genetically modified plants and seeds. She sees the Green Revolution as having two parts. "The first green revolution was public sector driven. It was driven by government agencies [that] controlled the research. In the case of the second Green Revolution, it is driven by Monsanto. It is a Monsanto-driven revolution. The second difference is that the first Green Revolution did have a hidden objective to sell more chemicals, but it's first objective was providing food; it was providing food security," said Shiva. "The second Green Revolution has nothing to do with food security. It is about returns to Monsanto's profits. That's all it's about."

Since 2001, agronomists Kiran Sakhari and Abdul Gayum have studied the growth of GM cotton by small farmers in India. Keeping track of yields and production costs, they publish a yearly report that compares GM cotton with conventional cotton. In 2006, the cotton crops were inundated with a disease to which GM crops seemed particularly susceptible.

"When we were doing our study from 2001," said Gayum, "we noted this disease in very few samples, in the *Bt* cotton only. As the time passed, the spread was seen more and more in *Bt* fields as well as some non-*Bt* fields."

The appearance of this new threat to cotton plants had never been seen before the introduction of GM cotton, and Gayum thinks it may be as a result of a negative, unforeseen reaction between the host plant and the gene that is carrying the insecticide. This is only one example of the many unknowns involved in genetic modification.

Sakhari added that while Monsanto advertising claims that using *Bt* cotton will reduce the pesticide reduction by 78 percent and increase the yields by 30 percent, this has not been borne out by farmers' experience. "A 2005 study of 87 villages in the state of Andhra Pradesh over a period of three years showed that … the performance of *Bt* cotton was erratic and plagued by numerous problems, including increased pests and disease, brittle stems, failure to germinate, drought damage, smaller cotton bolls, increased labor requirements, poorer quality and a shorter harvest season. Some farmers complained 'that they were not able to grow other crops after *Bt* because it had infected their soil' … The overall average return to *Bt* cotton farmers over three years was 60% less."[14]

Similar poor yield results were found with GM soybean. "A range of industry-, university-, and state-sponsored surveys summarized by Benbrook and Martinez-Ghersa et al. showed that RR [RoundUp Ready] soybean yields averaged 5-10 percent less than conventional soybeans."[15]

As well, a recent study conducted in Kansas confirms the experience of farmers. The study found that, "genetic modification actually cuts the productivity of crops … undermining repeated claims that a switch to the controversial technology is needed to solve the growing world food crisis. The study – carried out over the past three years at the University of Kansas in the US grain belt

---

[14] Jeffery Smith, *Genetic Roulette* (Fairfield: Yes! Books, 2007) 241.
[15] Smith, 241.

– has found that GM soya produces about 10 per cent less food than its conventional equivalent, contradicting assertions by advocates of the technology that it increases yields."[16] Also, when Professor Robert Watson (chief scientist of the study and director of the World Bank's International Assessment of Agricultural Knowledge, Science and Technology) was asked if GMOs are the answer to world hunger, he said, "The simple answer is no."[17]

In conclusion, "the general absence of a yield benefit discounts the notion that biotech is necessary to 'feed the world,' or could somehow do better than conventional plant breeding is already doing."[18] In fact, traditional farmers have almost always found ways to deal with the environmental vagaries of their specific regions. "By relying on [a variety of] traditional crops, we have coped with years of drought and never faced hunger. We have adapted our crops to local conditions and grow our food on marginal soils with no irrigated water," said a small farmer from India.[19]

Instrument of Oppression

Third World countries may be poor from North American standards, but they possess a wealth of agricultural knowledge specific to their needs – knowledge that has been passed down through the centuries. It was this exact type of knowledge that the Irish did not have when the British imported the potato to Ireland. The British had their tenant farmers in Ireland grow wheat for import to Britain and potatoes for the Irish to subsist on. This arrangement

---

[16] Geoffrey Lean, "Exposed : The Great GM Crop Myth," *Independent UK*, 20 Apr. 2008 ‹http://www.saynotogmos.org/ud2008/uapr08b.php›.
[17] Lean.
[18] Smith, 240.
[19] "Farmers, scientists and policy makers from the developing world say no!" Council of Canadians media release, 5 Mar. 2005 ‹www.canadians.canadians.org/media/food/2005/09-Mar-05.html›.

worked very well for a number of years because the spud is a very nutritious tuber capable of sustaining life indefinitely. It's also an unlikely instrument of oppression, according to historian Salaman.

"The potato can, and generally does, play a twofold part: that of a nutritious food and that of a weapon ready forged for the exploitation of a weaker group in a mixed society,"[20] wrote Salaman. This happened in Ireland under British rule, in Peru when the Spanish invaded, and of course when North America was colonized. "If for any reason," Salaman writes, "good or bad, conscious or otherwise, it is in the interests of one economically stronger group to coerce another, then in the absence of political, legal or moral restraint, that task is enormously facilitated when the weaker group can either be persuaded or forced to adopt some simple, cheaply produced food as the mainstay of its subsistence. Experience shows that this course inevitably results in a lower standard of living. The lower that standard, the easier is the task of exploitation and the nearer will the status of the weaker class approximate to serfdom."[21]

By the time the spud was firmly embedded in the diets of the tenant farmers of Ireland, the potato plant was weakened by repeated cloning and so was susceptible to blight – a disease that continues to be the most deadly of diseases for the potato worldwide.

Late Blight

In order to understand this disease, it's worth taking a moment to describe how potatoes propagate. Potatoes can be grown by cloning or by seed. Cloning is a process of taking a potato from your yield, and planting that for the next year's crop, either whole or cut up. If you've ever

[20] Salaman, 600.

[21] Salaman, 600.

driven through the countryside and seen signs that say "Potato Seed," what they're actually referring to are clones, not real seed. Potatoes do have seeds, but rarely are spuds grown from them because you can't be sure what type of potato will grow from them. For example, if you have a Russet potato plant that has gone to seed, you collect that seed. Then when you plant the seed, you could end up with any type or colour of potato. The possibilities in one tiny seed are astounding.

Raymond Loo, PEI organic farmer, explained how many people don't realize that apples trees aren't planted by seed either, but rather grafted.

"If you took a McIntosh and planted a seed from that McIntosh, you won't get a McIntosh apple. Potatoes are the same, you see. Like Russet Burbank, it's picked out of the ground and cut up ever since. You see it stays the same. That's why disease pressure on potatoes is such a big problem," said Raymond.

In the mid-1800s in Ireland, enough about how to grow the potato wasn't known. They didn't realize that the ability of the potato to resist disease decreased with each generation of cloning until the potato became so weak it couldn't resist the late blight at all, and eventually it succumbed. An analogy to understand this process might be photocopying. If you keep making a photocopy of a photocopy, eventually you will end up with a black mess. That's what late blight did to the spuds during the Irish Potato Famines – turned them into black mush. The disease that wiped out the spud, which led to the deaths of so many in the 1800s, continues to be a dangerous threat to potatoes worldwide.

Developing a blight-proof potato has been a goal of Green Revolution scientists as well as seed corporations. So far, they have not managed to do what organic farmers in

Peru and elsewhere have managed to do by working with the environment and using accumulated knowledge.

In 1999, Monsanto had two varieties of GM potatoes approved in Canada, New Leaf Y and New Leaf Plus. They "are genetically engineered to repel two separate potato viruses [and] to resist the Colorado potato beetle, a common pest."[22] These GM potatoes are not sold commercially; however, because consumers made it clear they did not want to eat genetically modified potatoes.

"Genetically modified potatoes are getting a firm thumbs down from North America. McCain Foods announced this week that it would not use genetically modified crops in any of its products. The move is a response to consumer demand."[23]

More recently, BASF: The Chemical Company based in Germany, has developed a GM potato designed to resist blight. It was made by inserting a gene from a wild Mexican potato that has developed resistance to blight naturally. Regardless of the expensive genetic tinkering, chemicals to prevent blight still need to be applied on these GM potatoes.

"Our hypothesis is it will dramatically reduce blight sprays," said a BASF spokesperson, "but to get the most out of the technology we need to use an integrated management strategy. At a minimum the trait should significantly increase spray intervals"[24] – but not *eliminate* the need for pesticides.

---

[22] Pauline Tam, "Government Fast-tracked Monsanto's GM Potatoes," *Ottawa Citizen,* 30 Nov. 99 ‹www.mindfully.org/GE/Monsanto-Potatoe-Fast-Tracked.htm›.

[23] "McCain Blows Cool on GM Potatoes," *Farmers Weekly,* 3 Dec. 99 ‹www.mindfully.org/GE/McCain-Blows-Cool-Potatoes.htm›.

[24] Mike Abram, "What does GM potatoes mean for future blight control?" *Farmers Weekly*, 7 Dec. 2006 ‹www.fwi.co.uk/Articles/2006/12/07/100127/what-does-gm-potatoes-mean-for-future-blight-control.html›.

Natural Solutions

So, what can be the use of such a potato? Despite all the science that went into it, it still needs chemicals as a protection against the blight. It seems to me that if you already have a system in place that reduces or eliminates the need for chemicals, why go the scientific route at all? But is there such a system in place that can help feed the world? Yes, and more about these methods can be found at the Rodale Institute in the States, for example, or by reading the books of Raoul Robinson.[25] Organic farming techniques like theirs are mirrored in long-standing agricultural practices around the world. Just like the farmer from India who said they had developed a way of growing food within the existing environment, small farmers have also found ways to cope with the late blight virus. For example, the Quechua are experiencing blight problems right now, and what they are doing is moving to areas that are less wet and warm, as this exacerbates the spread of late blight.

"We are moving our crops higher up the mountain," a Quechua farmer said. "They are healthier, and we get a better yield at the higher altitude."[26]

The Quechua can do this because they are part of a community of over 8,000 villagers that share a vast mountainous region. For other farmers who don't have the luxury of relocating, there's the option of ensuring the health and resistance of the potato plant.Some potatoes, like the Mexican one used by the German chemical company BASF, are naturally resistant. Breeding those potatoes with others to create a stronger, natural potato is a low cost and highly efficient way to deal with the problem of blight. But this type of breeding needs to be on-going to be effective.

[25] See appendix.

[26] Joanne Silberner, "In Highland Peru, a Culture Confronts Blight," ‹www.npr.org/templates/story/story.php?storyId=87811933&ps=bb1›.

The farmer must stay involved and in tune with his/her crops to ward off ever evolving blight infections.

"You can get [a potato] that is resistant, but the disease changes. The potato doesn't," said Raymond.

So far there is no scientific magic bullet to eliminate the problem of late blight. But there may not be a need to do so since farmers have been able to stay ahead of the problem through resistance breeding[27] and using knowledge gained over the centuries. To do this a farmer needs to remain small enough to stay in touch with the needs of his plants, something large factory farmers are unable to do.

Quechua farmers "have to adapt their crops to those conditions, and they look for how to create conditions in every little niche," Argumedo says. "So the diversity of crops they created is a response to the chaos of the system."[28]

There's a similarity in approach between what Argumedo details here and the *clachans*, the Irish land-sharing approach described by Kevin Whelan. Because of their communal approach to life, every family in the *clachan* received a portion of each type of soil available so that everyone had access to earth that was suited for a variety of crops. There was a fairness of distribution that ensured the wealth was shared by each family.

When the British colonized Ireland, they imposed a hierarchical system of doing things that eventually obliterated the *clachan* system, according to Whelan, as well as the Irish system of land and knowledge sharing, a system similar to what the Quechua and other native communities still retain today. Perhaps if the British mono-crop system (that was based on the values of individualism) had not replaced the communal Irish farming system, the Irish might

---

[27] Raoul Robinson, *Return to Resistance: Breeding Crops to Reduce Pesticide Dependence* (Davis, Agaccess. 1996).

[28] Silberner.

have been better able to collectively cope with occurrences of blight – and with much less devastating consequences.

Island Sunshine

As well, with the *clachan* system, perhaps the Irish might have stumbled upon the same discovery as did the Loo brothers of Prince Edward Island. The Loos' story starts back in the seventies and is one that demonstrates how important access to a wide variety of seeds and plants, unencumbered by corporate control, is for human creativity and ingenuity to flourish.

Two of Raymond Loo's uncles found some potatoes in their strawberry patch and saved the spud seed balls. Then they began growing them under grow lights in their basement, keeping records on a cornflakes box, "and that's when they figured out that every potato is different," said Raymond. That knowledge came as a big surprise to these potato farmers used to growing from clones they purchased, or from potatoes they saved from each harvest.

Potatoes have been growing through this cloning method for so long most don't know that, firstly, potato have seeds, and secondly, that you can grow potatoes from seed. Once the Loos understood how potato seed worked, they started experimenting with cross-pollination of different varieties. Because potatoes don't reproduce sexually as easily as some other plants, getting two varieties to cross breed has to be done manually. The flower of one variety is mingled with that of a second variety and the resulting offspring is a combination of the two types of potatoes. That's how the Loo family's Island Sunshine potato variety came into being. Specifically, the Loos crossed a Binchey with a Red Irene. From the resulting seed ball, seeds were grown and one of them became Island Sunshine, which is now propagated via cloning and seed potatoes.

By the eighties, what began as a hobby blossomed into "becoming quite a time-consuming almost obsession. [Gerrit, Raymond's father,] was going everyday, all summer, out to Afton Road ... a little field in Riverdale, twenty minutes from here. My uncle had land; the field was completely surrounded by woods, big lot of woods, and no potatoes in that area. Very hilly. There had been no potatoes around for a long long time. So, he started growing all these different varieties out there in that field."

Raymond recalls that it was a very impressive sight to see such an expanse of many different new varieties growing for the first time in that soil. It was profound to him what "two retired farmers ... who didn't have any agricultural degrees or anything" had accomplished.

Getting Island Sunshine certified took many years of dogged determination, however. Because they were organic potatoes of a variety created by the Loos, the authorities were highly suspicious.

What really surprised the Loos was how little the authorities back then really knew about how potatoes were propagated and how new varieties could be created. Even some highly educated people working in the agricultural sector of the government didn't know that potatoes could be grown by seed, recalled Raymond.

During the inspection of the Loo fields, government authorities were very skeptical of these strange potato varieties they found the Loos growing. They suspected that the Loos were illegally bringing in unapproved varieties over from Europe.

"They didn't believe that [my father and uncles] were taking these seed balls and getting all these varieties," said Raymond. "So when Dad started talking about seed, true seed, they were talking about potato clones. They were talking two different languages. And most people, even these guys [who] were doctors and highly educated, they had no idea how potato seed works."

Even neighbouring farmers displayed a lack of understanding about organic farming, according to Raymond. Trying to get government support for a new project, Raymond had to get letters from his neighbours. One conventional (chemical) farmer wrote a letter that said what Raymond was proposing to do on his own farm wasn't any worse than what he was already doing, recalled Raymond.

Working against this lack of knowledge within the government and the neighbouring farms made the process of certification an arduous one, but Gerrit persevered. Raymond recalled that local farmers would even turn their backs on the Loos in meetings – so disapproving were they of organic farming, despite the little they really knew about it.

"Dad was the one who went to all the meetings and set everything up and, you know, Dad being Dad, he wasn't deterred easily. Nobody was going to intimidate him whether by having a big title or a big name. That just wasn't an issue for him. I think some people probably would have given up, but that wasn't something Dad did. He just didn't give up, and you know when he was in the hospital bed with leukemia, dying, the day before he died, he was still saying if he had a heart transplant, he'd be all right ... So, there was never ever a time he gave up."

The final push to register their blight-resistant potato, Island Sunshine, came when one day, a letter from Monsanto arrived asking for some of their potatoes so they could study them.

"That would be about 1991/2," said Raymond. The letter said "that they would like to have a sample of Island Sunshine so they could isolate the genes that are making it resistant to blight."

Given that Monsanto's business is to patent genes and then demand royalties for what they have genetically modified, this was a very significant turn of events in the lives of a small farming family from this tiny Atlantic province.

"Dad was astounded about how Monsanto heard of us. You know it almost blew us away. Here we are little farmers in PEI."

Because of this interest from Monsanto, the Loo family applied for plant breeders' rights over Island Sunshine, which they received in 1995. That way, they'd have legal recourse if anyone took samples of their potatoes to develop a genetically modified version of it. But as you'll see in the next chapters, defending legal rights can be a costly thing both financially and emotionally.

Not So Simple

Now, you may be thinking, if the yields of GM crops are poor, pesticide use has increased, and GMOs aren't the answer to world hunger after all, then why don't the farmers just return to growing with natural seed? Apparently, that is not an option for many farmers in India because natural seed is not available anymore, in most areas. In Paraguay, the soybean seed has been irrevocably contaminated with GM seed, and now Mexico's ancient corn varieties are threatened by the infiltration of GM corn. This is a shocking development for the world's largest genetic pool of corn.

For First World farmers and consumers alike, the future doesn't look much better. According to Professor James of the Rowett Institute, a nutritional research facility in the UK, "We doubt it will be possible to keep all GM foods completely separate in 10 years' time."[29] Alarmingly,

---

[29] Philip James, "Memorandum submitted by the Rowett Research Institute, Aberdeen," *Minutes of Evidence*, UK Parliament, March 1999

this was said in 1999, but it didn't take long for James' apocryphal words to become reality. In 2000, hundreds of products had to be recalled worldwide because an American brand of GM corn, Starlink, manufactured by Aventis SA, a French drug company, had infiltrated the human food system.[30] This GM corn was destined for animal feed only, but had somehow found its way into dozens of consumer products. The next year, Japan repeatedly found traces of unapproved GM potato in their snack foods that used ingredients from the United States.[31] Though these imports came with certificates stating they were GMO-fee, the Japanese government found the potato snacks were contaminated with NewLeaf Y and NewLeaf Plus – both products of Monsanto. The cost of recalling these foods was enormous.

For small farmers, however, the cost of GM seed in India is often four times higher than natural seed was. Plus, along with GM cotton seed, farmers must buy the Monsanto-produced pesticide designed specifically for the *Bt* cotton. This pesticide is also very expensive, and even more so now that more of it is needed to control insect infestations. As a result, farmers are often forced to borrow cash from moneylenders, resulting in bankruptcy for the farmer if crops fail. In fact, a study of the Andhra Pradesh region of India found that "71% of farmers who used *Bt* cotton ended up with financial losses. (Similarly, about 70% of the 4,438 farmers growing *Bt* cotton in Indonesia were

‹www.publications.parliament.uk/pa/cm199899/cmselect/cmsctech/286/9030813.htm›.

[30] Julie Vorman, "Starlink recall climb to 300 different items," Reuters News Service, 1 Nov 2000 ‹www.organicconsumers.org/ge/starlink300.cfm›

[31] "Morinaga recalls potato snack over GM concerns," Reuters News Service, 12 July 2001 ‹www.planetark.org/dailynewsstory/.cfm?newsid=11530›.

unable to repay their credit after the first year of planting.)"[32]

Faced with insurmountable debt, many farmers in India subsequently kill themselves. In the Warangal district of India, *Bt* suicides occurred as often as three times a day once GM cotton was introduced in 2005. In the first year that *Bt* cotton was introduced, there were 600 suicides, according to farm activist, Kishor Tiwari. "Within the first six months of the next year, there were 682 suicides." In a terrible irony, many of those farmers killed themselves by drinking Monsanto pesticide.

That people have been driven to such undignified ends because of GM science that was supposed to improve their lives underscores how profoundly genetic engineering disturbs the relationships amongst people as well as between people and their environment. It's a world that has been transformed by the bottom line, a world disfigured by profit margins. Having work that we can be proud of, feeling like we have some control over our lives or at least are able to take care of our families is crucial for our sense of self-worth. Factory farming and GM technology is wreaking havoc with the individual's ability to work in harmony with the environment. The type of relationship with the environment that the Irish had in their *clachans* and that some Native agricultural communities still have today may seem nostalgic. However, it is in fact a foundation upon which societies can create a community they are proud of and in which they can thrive and prosper.

"We are indeed *homo faber*, a species content only if we have work. But the jobs we do don't always fulfill our dreams. And when our work lacks dignity, so do we," said Matthew Fox in *The Reinvention of Work*.

---

[32] Smith, 241.

The Independent Life

Perhaps it is because of work that lacks dignity that there are still those farmers who remain independent like John. "I would never consider working for the big man ...They're a business out to make money off your back," said John Shepherd. "If I was having to sell to them, I wouldn't grow spuds, because all they do is dictate the terms, and they just keep a wee carrot dangling in the front of you. Every time you think you're getting near the carrot, they move it again. They just keep moving the goal post."

Raymond shares the same desire to work for himself. "I'm not scared of competition," he said. "That's, of course, something I grew up with. We didn't have any money. Dad was in a bad car accident in 1970, but we always had enough food. We have our own beef, pork, chicken, eggs, vegetables. We make our own preserves, our own jams – not that we don't buy groceries, but I like the quality we can do, and it gives us independence. I like independence. That's why I like working for myself."

Both Raymond and John enjoy an independence that's hard won, but the vital contribution farmers make in society is often as overlooked as the potato is.

"Do you feel proud that you supply just to Northern Ireland?" Fred asked of John.

"I don't really think much about it. I just grow spuds, and that's my life, and that's it ... We had an awful lot of wee small farms here. But they're all just melted [away]. There were two small farms. Now there's only one of them. The other farm is no more.

"What happened to the land? Did you take it over?" asked researcher Fred.

"A whole lot of it is planting trees," said John. A current trend in Northern Ireland had been to pay farmers not to grow food but rather plant trees. "All you get out of this government now is that it's cheaper to buy in. Import everything. Well, at times you have to buy imported stuff, but if you can, keep it at home as best you can. You're keeping your money at home. What's Tony Blair going to do twenty years down the line?"

"Given this government, do you feel there'll be a place for you down the road?" asked Fred.

"No," said John, but he's not without some hope for a distant future. "I honestly feel that the tenor of public opinion has turned full circle. People will get fed up, and they'll want the goods, they'll want their own again, [and] if the people are prepared to go and get it at a farm, it's cheaper than buying it out of the supermarket."

The niches both John Shepherd and Raymond Loo have managed to proudly carve out for themselves are tempered by the walls of industrialized farming and GMOs closing in around them. However modest, their farms reflect a lifetime of hard work and dedication. In a way, like the potato god, they must have many strong teeth with which to create their farming success.

Despite the fact that most farmers around the globe share the desire to be independent, self-reliant, and do work that gives them dignity, there is a sharp contrast in the situation between farmers in developing nations and those in Western nations. Food and trade policy analyst, Devinder Sharma, put these issues into perspective. "The number of farms in the UK today is less than 0.7 percent of the population," said Sharma in a 2001 interview. "In India it is 70 percent of the population that are directly or indirectly engaged in agriculture. Today there are something like 550 million farmers in India. What is good for the North is not good for the South."[33]

So, what is at stake for small farmers in poorer countries is proportionately greater than in the Western

---

[33] Devinder Sharma, interview, Norfolk, UK, 2001 ‹members.tripod.com /~ngin/feedtheworld.htm›.

world, though no less significant to the individuals involved, of course. If a small farmer in the Western world were to lose his farm, that would be tragic. However, there are other options, other jobs to be had, unemployment, or even welfare for that 0.7 percent of the population. If 550 million people in India were to lose their livelihoods, the devastation would be unimaginable.

In North America, we take for granted that if you work hard, your success will naturally follow, but that belief doesn't bear fruit for small farmers in places where GMOs have come to stay, as in India and Paraguay. In such places, the walls have already closed in on many independent farmers. For them, mono-crop farming and being surrounded by genetically modified crops that taint their natural crops is not a situation they thrive in with dignity. In the opinion of farmers' activist Jorges Galhario from Paraguay, "It brings death, poverty and illness as well as destruction of the natural resources that help us live."[34]

Ironically, this destruction is wrought by a few people far removed from the effects of the technology they are dispersing. "Ten people sitting in a board room decide what is good for [this other] society [supposedly] in the name of food security," said Devinder Sharma. "They are not actually interested in food security; they are actually interested in profit security for their companies."

So What?

After all this, you might say, *Well, it's too bad that GM technology is contaminating everything, but if we can't stop it, as least we still have food to eat, right?* Not necessarily.

Because GM technology is so expensive, it pushes up the price of food. For the millions in India, that means many people can't afford to buy all of the food they need.

[34] *The World According to Monsanto*

"In our country," said policy and trade analyst, Devinder Sharma, "there is 50 million tons of surplus and 250 million people go to bed hungry every night because these people cannot buy this food. Biotechnology in fact adds to this problem of hunger and malnutrition. With more and more biotechnology we're going to end up in what is called the paradox of plenty which means our buffer stocks will go up, our grain silos will be bursting and people will be outside the reach of this food and agriculture. That is a very strange and dangerous phenomenon but somehow the policy makers and the scientific community is not aware of all this or is certainly turning a blind eye."[35]

This situation bears eerie resemblance to those in ancient Peru and in Famine Ireland. In Peru during the Spanish colonization of the 1500s, as mentioned earlier, many slaves working in the silver mines had to buy back their own dried potato, or starve. In Ireland of the 1800s, there was what some call a "false famine," because while the potato crops were devastated by late blight, those same tenant farmers were also growing grain for their landlords back in England. That grain was not used to help the starving Irish, but rather shipped, as usual, into the coffers of the British landowners. Salaman's contention that the potato can be used as an instrument of oppression in a society divided by class is just as relevant a statement today as when he made it back in the 1940s. So, that gap between urban and rural communities gets wider the more GM technology spreads. However, instead of upper and lower classes, that gap is characterized by biotech companies on the one hand and the rest of us on the other.

Added to this disparity is the gap between what is safe to eat and what isn't. Proponents of genetically modified foods assert that it is safe to eat because it has

---

[35] Devinder Sharma, interview, Norfolk, UK, 2001 ‹members.tripod.com /~ngin/feedtheworld.htm›.

been deemed "substantially equivalent." But this phrase was coined to describe GM food that appeared alike to that of natural food. Substantial equivalence *does not* mean nutritionally the same. Although genetically modified foods have been floating around in our food system since the early/mid 1990s, no food safety tests had been done on potatoes, or any other food destined for our tables, until eminent British lectinologist Dr. Arpad Pusztai entered the scene in the late 1990s. His discoveries would rock Europe and the UK in 1998/1999. Oddly, the shock waves barely made a ripple in North America.

Chapter Three – Magyar Maverick

Most people in North America have never heard of Arpad Pusztai, but in the UK in 1998, his name was splashed across the media for months in a potato scandal that changed his life forever.

Dr. Pusztai worked for the prestigious Rowett Institute. Located in Scotland, it's one of four publicly funded nutrition research institutes in the UK and an independent company with charitable status. It receives the majority of its funding from the UK government, but also receives a percentage from biotech firms such as Monsanto. However, Pusztai estimated Monsanto's contribution, in 1998, at less than one percent of the total Rowett budget.

Pusztai had worked at the Rowett as a nutritional scientist for decades, and at the time of the potato scandal, he was working, along with his wife, Dr. Susan Bardocz, on testing the nutritional safety of genetically modified potatoes. He was chosen for this work because of his long track record as a prolific scientist, and because his specialty, lectinology, was particularly relevant. "Dr Arpad Pusztai is a distinguished expert in lectins, i.e. plant proteins of complex and varied structure with a very broad range of actions," wrote Professor Philip James, the director of the Rowett at that time. "In plants, lectins seem to act by deterring pests: lectins are frequently found to damage the intestine of insects, nematodes, etc. Therefore they have become popular with plant molecular biologists, since the insertion of a lectin into a major crop may enhance its pest resistance and reduce the need for spraying with alternative pesticides. Dr Pusztai showed years ago, however, that some lectins, e.g. the phytohaemagglutinin lectin (PHA) obtained from red kidney beans can induce very marked intestinal damage to the mammalian gut."[36] So, having Dr. Pusztai test the food

[36] James, memo ‹www.publications.parliament.uk/pa/cm199899/cmselect/

safety of potatoes that were genetically modified with a lectin gene was the obvious choice.

In August of 1998, six months into his research, a brief television interview was arranged by Prof. James for Dr. Pusztai to discuss the preliminary results of his work. This was seen as a necessary thing in order to generate interest and possible corporate funding for this crucial food safety work.

"It is very important that we should remember that this business of me going on the programme was very much a part of the normal publicity," Dr. Pusztai said in the UK parliament, "sort of what you get nowadays, because you have to raise money—you have to raise money—and most of the reason why, eight years after I retired, I was there was because I raised a lot of money. It is as simple as that."[37]

What seemed like a routine bit of publicity for the Rowett turned out to be a public relations disaster. In the interests of both public safety and scientific integrity, Dr. Pusztai matter-of-factly reported that, so far, his research indicated that genetically modified potatoes were unsafe to eat.

In short, Dr. Pusztai's study found that, "the GM potatoes adversely affected virtually every organ system of young rats – with most changes found after just 10 days. Their brains, livers, and testicles were generally smaller, suggesting disruption of normal growth processes due to either malabsorption of nutrients or unknown toxins. White blood cells responded to a challenge more slowly, indicating immune system damage; and organs related to the immune system, including the thymus and the spleen, also showed changes. The animals had enlarged pancreases and

---

cmsctech/286/9030813.htm›.

[37] Arpad Pusztai, examination of witness, Committee on Science and Technology, UK Parliament, 8 Mar. 99 ‹www.publications.parliament.uk /pa/cm199899/cmselect/cmsctech/286/9030806.htm›.

intestines, and partial atrophy of the liver. And in all cases, the GM potato created proliferative cell growth in the stomach and small and large intestines; the lining was significantly thicker than controls."[38]

Dr. Stanley Ewen, Pusztai's partner in the study, stated that as a hospital pathologist, there were "significant differences"[39] between the rats fed natural potato and GM potato, and that this growth was cause for concern, because "although no tumors were detected, such growth can indicate a pre-cancerous condition."[40]

This was a great surprise to many people, including Pusztai himself. When the Rowett was hired by the Scottish government to test GM potatoes, there was an expectation that this would be a big part of the UK economy's immediate future. It was also assumed that the testing was just a rubberstamp affair, because the common belief at the time was that GM foods were "substantially equivalent." In other words, *If it looks like a potato and tastes like a potato, then it's a potato.* However, it does not mean it is safe to eat.

This erroneous conclusion stemmed in the main from the one and only published study, dated 1996, that said GM soybean fed to livestock was substantially equivalent. All the GM foods now in the system were approved on the basis of this single study, *which tested only one GM substance!*[41]

---

[38] Smith, 23.

[39] Stanley Ewen, "Memorandum submitted by Dr Stanley William Barclay Ewen, Department of Pathology, University of Aberdeen," UK parliament, Mar. 1999 ‹www.publications.parliament.uk/pa/cm199899/cmselect/cmsctech/286/9030804.htm›.

[40] Smith, 23.

[41] Hammond BG, Vicini JL, Hartnell GF, Naylor MW, Knight CD, Robinson EH, Fuchs RL and Padgette SR. , "The feeding value of soybeans fed to rats, chickens, catfish and dairy cattle is not altered by genetic incorporation of glyphosate tolerance," *Journal of Nutrition*, 126:717-727, 1996.

In a memorandum to the UK parliament, Prof. James and Dr. Chesson were critical of the way some scientists and others may jump to the conclusion that genetically modified foods are automatically safe because of being labeled substantially equivalent. Inadequate experiments are the basis for this incorrect assumption, they explained. Since it's difficult to get enough protein from one GM plant to test it, the commonly accepted practice in science is to use a bacterium. First, scientists take the foreign gene that the biotech company wants to modify the natural plant with. Then they insert it into a bacterium to see how the bacterium reacts to this foreign gene protein. But, James and Chesson pointed out in their memo that proteins express themselves differently depending on what hosts they're in – whether it's a bacterium or a plant. Therefore, "extrapolating from the tested behaviour of an isolated protein produced in a bacterium to predicting the behaviour of the same protein when it is an integral part of the transgenic plant is unsound and could lead to premature conclusions about safety."[42]

Though Pusztai's experiments had proved the assumption of nutritional substantial equivalence false, no one expected his findings to cause the media storm that it did. Caught off guard, Director James issued press releases that contained errors about Dr. Pusztai's study and had the effect of ruining Pusztai's reputation practically overnight. In the aftermath of Arpad's television appearance, official word from the Rowett Institute indicated that Dr. Pusztai, a scientist with decades of experience and hundreds of peer-reviewed articles, did not give the correct information. As a result, Arpad was quickly shoved into the background and told not to speak to the press.

---

[42] James, memo ‹www.publications.parliament.uk/pa/cm199899/cmselect /cmsctech/286/9030811.htm›.

"The interview went out at 8 p.m.," said Arpad, "and by 11 a.m., he [Director James] had relieved me of the communication between the media and the Institute because 'he was much more skillful, and I needed protection because I would not do quite so well.' So, in fact, from 11 a.m. onwards, my 'phone calls went up to the director's office."

This was all going on behind Arpad and Susan's back. As far as they were concerned, nothing out of the ordinary had happened: Philip James had asked Arpad to do an interview. He did it. It went well, and that was that. However, it wasn't until the next morning at work that they realized something was amiss.

"We found out simply by accident. The phone was quiet, and there were no phone calls," said Arpad. He later realized that his son, who lives in Edinburgh, was trying to get in touch with him, but none of the calls for either Dr. Pusztai or his wife, Dr. Bardocz, were put through to them.

"My emails were intercepted," continued Arpad, "and in all the press releases the contact at the end was always Professor James, not me." In fact, of the press releases sent out after his TV interview, Dr. Pusztai was shown only one. Except for that one, he wasn't given any others to verify before they were sent, as was the usual custom, "otherwise I would've been able to correct them," said Arpad.

Political & Corporate Pressure

As the drama played itself out over the next weeks and months, Arpad and Susan pieced the events together. The extent of their situation revealed itself gradually from what they read in the newspapers, and through how they were treated by their long-time colleagues.

"We know there were two telephone calls from Number 10 [Downing Street, the Prime Minister's office].

Whether it was Tony Blair himself or some of his advisors," Pusztai didn't know, but "all the secretaries, for example, Susan's secretary, had to man the telephones in the director's office. So, we'd been unofficially told about these telephone calls," said Arpad. Obviously the content of these calls we don't know. But we can put two and two together."

Dr. Stanley Ewen, Arpad's colleague and fellow researcher at the Rowett Institute, was in charge of evaluating the impact of GM potatoes on the rats used in the experiment. Speaking about the situation years later, he said, "I was extremely angry and very very concerned. It's like your whole world is disappearing under your feet. What's going on? Monday it was wonderful work. Tuesday it was rubbished."

Though Ewen was a partner in this study, it was Pusztai who took the brunt of the fallout.

"On Tuesday, the 11th, we parted company at 5 p.m. from the director's office with the deputy director being given the task of putting Susan's factual bits of information into a form that would be releasable," related Arpad. "That was the final thing. Nine o'clock the next morning, we're called and suspended. So something must have happened between 5 p.m. on Tuesday afternoon and 9 a.m. on Wednesday morning."

Although Ewen had his suspicions, he wasn't sure until some time after Arpad was dismissed and their research team dismantled.

"I was at a dinner dance and next to me at the top table was someone from the Rowett. I said, 'Isn't it awful what's happening to Arpad?' 'Yes,' he said, 'and did you not know that there were not one but two phone calls from Downing Street to the Director?' And then of course, I saw clearly what was happening," said Ewen. "This was something supranational, if you like. Some pressure being put on Tony Blair's office to stop this work because it was perceived by the Americans to be harming their industrial base, the biotech industry, in other words."

Despite the talk about phone calls from the Prime Minister's office in response to this television interview, Professor James testified in the House of Lords that he did not take any such calls and that his handling of the situation was "totally free from any influence, at any level, whether it is political, industrial, please name it."[43]

Though it's hard to know what goes on behind closed doors, it is known that Monsanto did give research money to the Rowett. It's also known that a representative from the biotech giant, Dan Verakis, was in the room at the time of Pusztai's interview where the unwelcome news that this type of GM potato was not safe was revealed. As well, it's not unlikely that Monsanto's position as one of the leading biotech companies in the world might be affected by such a revelation. In fact, in his capacity as spokesperson for Monsanto, Verakis had to deal with the subsequent onslaught of negative press and criticism from other biotech companies as well as consumer groups.

Gagged

When Arpad gave his short TV interview about his research on the food safety of genetically modified potatoes, it was a warning that the world desperately needed to hear. "As a scientist actively working in the field, I find that it's very, very unfair to use our fellow citizens as guinea pigs."[44] But that message was obscured by media reports that painted Dr. Pusztai as incompetent. What made matters worse was that Pusztai was unable to redress the errors made in the

---

[43] James, examination of witness, Committee on Science and Technology, UK Parliament, 8 Mar. 99 ‹ www.publications.parliament.uk /pa/cm199899/cmselect/cmsctech/286/9030816.htm›.

[44] *World in Action*, BBC Television broadcast, 10 Aug. 1998.

press over a six-month period because of the non-disclosure agreement he had signed with the Rowett Institute. This agreement is one that every scientist working at the Rowett is required to sign, and it's quite common to see such agreements in places where research is conducted, said Arpad.

"All state-sponsored research institutes have the very same contract, which each scientist will have to sign," explained Arpad, "and most of the time there is no problem with it. There's no problem, because it's in their interest for us to speak. But when there's this political pressure to gag someone, then it's very useful that they can just refer to the contract."

Susan pointed out that this contract, which she also signed, was a contract for life. Arpad emphasized that "it's not like the Official Secrets Act, which has a time limit on it. *This is forever.*"

It seemed that Arpad and Susan would have to endure the misrepresentation and vilification without recourse. Because of the non-disclosure contract, they could do or say nothing in their defense for as long as they lived. This created a very stressful situation for Susan and especially Arpad who could only stand by and watch as his reputation was systematically ruined.

"In the beginning, the Institute said it was all my fantasy," said Arpad. "My director said, 'He's never done this work.' And of course, he didn't give me permission to deny what he just said. After a time, the media believed that. In fact, they were sending around these stories to the media – the newspapers and radio and television – that 'poor old Arpad is very sorry to have said all these things. He now realizes how much damage he caused with his careless talk, and therefore he's very apologetic about it.' It was nothing of the sort, but I couldn't say this wasn't so."

Stanley Ewen, in his memo to the House of Lords, stated for the record that he felt the treatment Pusztai had to endure was terrible. "His friends and colleagues felt a real sense of outrage that Dr Pusztai," wrote Ewen, "a Hungarian refugee from KGB dominated Hungary in 1956, had been treated in this heavy handed manner."[45]

Adding to the stress were the reporters that pressed for information. Not surprisingly, Dr. Pusztai would have happily talked to the reporters. He has strong views on what science is and what a scientist's responsibilities are to himself and to the public. This is, in part, what set off the chain of events that would affect his life so greatly. Because the Rowett Institute is a publicly funded institution, it can be seen as a business answerable and in service to the public. This was certainly how Dr. Pusztai saw it and why he revealed his findings during that fateful television interview.

"You see, our money came from the public. So I was a public servant. That's why, in the first place, I agreed to that very short interview."

Despite what scientists and politicians in the aftermath of the scandal may have thought of him, Pusztai's belief in himself had never wavered. "The money came from the British public, and my warning had to be directed to the British public. So I couldn't care less what [scientists or politicians] think of me," he said.

As well, when asked in the House of Lords if he had any regrets about appearing on the programme, Arpad said, "No, because, in one sense, what I achieved is that we are all sitting here and talking about it."[46]

---

[45] Ewen, memo, ‹www.publications.parliament.uk/pa/cm199899/cmselect/cmsctech/286/9030804.htm›.

[46] Pusztai, witness examination ‹www.publications.parliament.uk/pa/cm199899/cmselect/cmsctech/286/9030807.htm›.

Mounting Stress

Because Dr. Pusztai saw it as his duty to inform the public, he would have been more than willing to break his non-disclosure agreement and speak to the media – but not without support. He felt that the burden of breaking his contract with the Rowett could not be borne on his shoulders alone. He describes himself as a man of modest means and, while comfortable, was certainly not in the position of being able to fund a potentially long and expensive legal battle between himself and the largest, most reputable scientific research institute in the British Isles. That an organization like Greenpeace did not step up to help was a situation which puzzled Pusztai.

Ostracized by most of his peers, painted as a doddering fool in the press, relieved of his duties, Pusztai's life had been turned upside down. Despite this humiliation, Arpad continued his work from home. (Though he had been relieved of his duties, his employment had not yet been terminated.) Much of Arpad's paperwork at the Rowett had been confiscated until the internal inquiry was conducted. In the meantime, he worked on other projects, as ironically, the potato research was only a small part of his overall research.

Unlike Arpad, his wife, Dr. Bardocz, continued to work at the Rowett. Only in her late forties at the time, Susan fought to keep her position and refused early retirement. Contractually, the Rowett had little choice but to let her stay on, she said. Her work was extremely curtailed, however, and her 'phone calls continued to be rerouted, as they had been since the day after the television broadcast.

At the Institute, Susan was shunned by her colleagues. "They would suddenly stop talking when I entered the common rooms," recalled Susan. The stress continued for both of them even though they tried to settle into their new routine. Each morning, Arpad would drive Susan to work and pick her up at the end of the day. During the day, he continued his work at home.

"The whole business was very unsatisfactory because Susan still had to go in," recalled Arpad. "I knew she was in this very hostile environment. She was not allowed to do any work. So she had to just sort of bide her time and put together the results of her experiments and all that. I used to go and pick her up by car around 4 o'clock. Our home is about six miles from the Institute. I was working on a scientific paper at home, looking at the screen of the computer and suddenly, it was a very funny feeling.

"By that time, it was about half past three, so it was coming up to – I would say by about 5 to 10 minutes more, I would've been in the car and driven … It really was a funny feeling. It's obvious that the blood supply to my brain was temporarily restricted, and the computer screen started to turn and the whole room started to go. You know, you suddenly lose all sense of reality. You don't know where you are. You look at yourself, more or less, from the outside. I tried to stand, and I just collapsed. The point that really frightened us was that in about five minutes' time I would have been sitting in the car, and if [the attack] happened while I was driving, that would have been very serious.

"I don't know how long I was down on the floor," said Arpad, "but as I was regaining consciousness, I managed to get to the phone and there was a code. I just pressed the button, and that rang in my wife's office. I was obviously very incoherent. So they were there within 10 minutes. Her secretary drove her home, and that was it. By the time they got home, I was reasonably coherent, and I managed to sit in one of the armchairs, and they called doctors. That's the reason I'm saying it was a mild heart attack because I survived it. And in fact, the doctor stayed with me for about an hour and half, chatting. He obviously wanted to make sure I was all right. Now I don't know whether the sudden surge of blood pressure – I don't know. I just remember the beginning of it starting. I remember this funny feeling of looking at myself from the outside. And I certainly don't want any repetition of it."

After the heart attack, working at the Rowett was even more stressful for Susan.

"This was about six, eight weeks after [the August broadcast]," recalled Arpad. "You see, unfortunately what Susan had to endure, perhaps it was even worse for her. She was only indirectly involved but she had to suffer all the consequences, too, and she had concern for me also. I'm twenty years older than she is. She didn't want me to be left alone. We didn't know if it would be repeated or not. But I think, by that time, mentally, I was over the worst."

Finding a Way

The heart incident brought home to Dr. Pusztai just how much he had been affected by the events. Not only did he lose his job, but he had to endure the world, and most importantly his peers, thinking he was a poor scientist.

"When some people in authority stood up and said, 'Well, this man is just talking about experiments that never existed,' normally, of course, you could sue these people for defamation of character. But I couldn't because I had to get the permission of my director to do it. So, it was really very frustrating.

"But one of my, let's say, better points is that in everything, I do look for the positive way out. Now what I came up with was very simple. By that time, we knew we would have to be given back the data because I had the right to reply. So, I took that and turned it to my advantage." Though it seemed hopeless because of the non-disclosure agreement, Arpad soon stumbled upon something that offered a possible way around an impossible situation.

When the Rowett had relieved him of his duties, Arpad had a lot of time on his hands. Rather than getting rid of an inconvenient problem, ironically the Rowett only provided Arpad with the time to find a loophole that would eventually vindicate the scientist and his work. Had he been occupied with his regular, full-time duties, he probably would not have had the time to slog through the huge document. Arpad knew that he would eventually have his say, somehow, but it was contingent on getting his data back from the Rowett. During the wait for the data and his convalescence, Arpad read the fine print of his contract, a prodigiously long document replete with baroque legalese.

While scouring through the contract, he found that his superior, Prof. James, was right in saying that Arpad could not speak to the public without prior permission from the Institute. But, Arpad also discovered that he could talk about and share his findings with his scientific colleagues, provided that nothing was published.

"My contract, true, did say what Professor James was telling me, forbidding me to give interviews, and so on, but what he could not do is stop confidential communication and exchange with other scientists, providing it was not published. So, I took from this – and I'm quite right, in this respect – that if other scientists asked me questions about what happened – What is my side of the story? What are the relevant facts? What were the results? – I could tell them providing they don't publish it."

Armed with this knowledge, Susan and he began putting together a document detailing his experiments. The process took several weeks since he had to wait to receive all his papers from the Rowett Institute, which came piecemeal. Once he had all his documents, Dr. Pusztai put the report together and sent it to his fellow lectinologists.

To all his colleagues who had asked questions, Arpad wrote a letter, in effect saying, "Look I'm willing to give you the official view and my response to it providing you do not publish it and that you give me your views as a referee."

Arpad sent the official audit committee's report, all 79 pages of it, along with his own data. Eventually, the two people in charge of the Society of Lectinologists wrote a memo that was posted on their web site. "What it said really," said Arpad, "was that 'we looked at the results, and as referees we can say, quite simply, the facts stand up. Although the results are preliminary, that doesn't mean that they are no good. So, we think that Dr. Pusztai was fully justified in saying what he did say in that interview.' So, that was one point. Scientifically that exonerated me, at least in the minds of those 24 people."

Support from his Peers

The Society of Lectinologists met in December of 1998 in Lunden, Sweden, where Arpad's colleague and potato study partner, Stanley Ewen, presented the facts at the meeting. "The abstract of that had actually got on to the web site of the International Lectin Society," said Arpad, "and a reporter from the *Daily Mirror*, I think, dug it out. So, he 'phoned me in the beginning of January, and [because of the non-disclosure agreement], I couldn't say anything except 'you've read it.'"

Though an article had finally appeared in the mainstream press that countered all the others, it still didn't release Arpad from his non-disclosure agreement, and there was no telling if this one article would stem the tide against him and his research. What happened next ushered in a scene that would be worthy of a Hollywood spy thriller climax.

When the article was printed in the daily paper, "the headlines read 'Pusztai's work has been vindicated' and this stirred the whole thing up again ... Both Prof. James and the head of my division had been officially invited [to contribute their opinions for the article], but they never even replied," said Arpad.

It had been five months since the television interview, and it was more than likely that people's attention was on other things by then. However, Arpad's research struck a nerve in British society. The Mad Cow devastation had already sensitized the nation to the dangers of ignorance, and there was an immediate response from all quarters, dubbed "The Great GM Debate." Therefore, the Science and Technology committee of the UK parliament convened an inquiry into the matter.

Shaken Foundations

Once the House of Lords (the highest authority in Britain) asked Arpad to speak, this overrode the Rowett's contractual gag, and Arpad could finally speak for himself.

Speaking in the House of Lords didn't restore Dr. Pusztai to his former life, but it provided an opportunity to have his side heard, and it went into the public record. "Because you see by then everybody, the whole media thought that Pusztai admitted that he was wrong, and he's very sorry for it and all that. People thought that it was genuinely like that, but now the whole [thing] erupted and now I could speak. So, it was a totally different ballgame."

The game had changed in Pusztai's favour, true, but even if his life could have picked up where it left off, things had changed irrevocably now. Arpad no longer worked at the Rowett. He was nearing the end of his career, anyway, but it was an inauspicious way to end his decades of scientific research. The events since his television interview and the subsequent heart attack had shaken the foundations of his personal life. As well, Dr. Pusztai's belief in his vocation had been profoundly affected.

"Now, I think it certainly has shaken my belief in science," said Arpad, "the way it is done today, and how much it is influenced by big business and politics. You see, for years, I kept hearing on TV this chief veterinary officer, the chief medical officer of the country – Jack Cunningham was the minister responsible – he kept saying on TV, 'There's no chance of this Mad Cow disease being transferred from cows to humans. It's unbelievable. It's impossible.' And then after three years, they had to admit that the impossible had happened."

In January of 1999 (four months after Pusztai's highly publicized television appearance and a month before the reporter's article about Arpad was printed), the British government "launched the "Bio-Wise" initiative, involving public investment of £13 million in Britain's biotechnology industry over the next four years." This announcement followed on the heels of two government committee reports which stated that GM crops did not endanger wildlife and "that the benefits of GM foods greatly outweighed the risks."[47]

Obviously Arpad's GM potato study had not been consulted. However, given his television appearance, news that cast doubt on the safety of GM foods could not have come at a worse time – what with a multi-million pound investment in biotechnology in the works.

Seeing programmes launched with huge budgets for biotechnology is a cause of great concern, because they are being given the green light before all the facts are in, according to Dr. Pusztai.

"In my case, Jack Cunningham stood up in Parliament and said that there was absolutely no evidence that what Dr. Pusztai says is true, and that the FDA in the United

---

[47] "The Great GM Food Debate - a survey of media coverage in the first half of 1999," Science and Technology Committee report, UK Parliament, May 2000 ‹www.parliament.uk/post/report138.pdf›.

States has investigated it in great detail, and they have come to the conclusion that there is no conceivable harm that could be associated with genetically modified potatoes. Now, [Cunningham] must have known that the FDA does not investigate any of these things, let alone our GM potatoes. They have no lab of their own, and when they are consenting to something being released, they *never* say that it is safe. What they say, in the letter – I have seen several – is that it is the company that says it's safe. Now, do you think the president of the Royal Society and Jack Cunningham are not fully in possession of the facts?"

If they weren't, they soon would be. In March 1999, Prof. James and Dr. Chesson of the Rowett Institute stated in their memo to the House of Lords: "We conclude that practices currently considered acceptable and promoted by the FDA are not rigorous enough for future use [of approving genetically modified foods]."

Anything that put a spanner into the works of partnerships between biotech companies and government would not be welcomed, though, because funding for scientists is very scarce. "In the case of the Royal Society,"[48]explained Dr. Pusztai, "78 percent of its funds comes from the United States, 32 percent [of that] comes from biotech and pharmaceutical companies. That does actually colour their judgment."

As recently as June 2008, there was still an old press release[49] from The Royal Society, dated 1999, posted on the

---

[48] "The Royal Society is the independent scientific academy of the UK and the Commonwealth dedicated to promoting excellence in science. The Society plays an influential role in national and international science policy and supports developments in science engineering and technology in a wide range of ways." ‹www.royalsociety.org›.

[49] "Genetically modified foods – Royal Society calls for Rational Debate," media release, Monsanto, Aug. 1998 ‹www.monsanto.co.uk/news/98/aug ust98/81398modified_foods.html›.

Monsanto UK web site. The Society's letter called for "rational debate" in the wake of the "prematurely released results" of the work of Drs. Pusztai and Ewen. Not surprisingly, there was no mention of these results later being substantiated.

"Now for the government," continued Arpad, "technically all governments are in favour [of genetically modified organisms] because what they see is that this is going to be the new industry that will generate all the funds. And these funds will come in; they don't see anything else on the horizon that is comparable. This is the huge new world of genetic manipulation. You know, they think that medicines based on your genetic make-up, sort of tailor-made to your genetic make up, will be the future. This huge amount of revenue will be generated out of this. But, you see, practically all of the government people have no training in science. So, they are relying on scientific advisors, [but] all of the scientific advisors are from the biotechnology and pharmaceutical industries."

Stanley Ewen stated in his memorandum to the UK parliament that while government advisory boards were made up of people who were qualified in their fields, the public didn't seem to be as well represented as businesses did. "The perceived orientation of the various committees seems distinctly producer based paying little attention to the needs of consumers with the result that intelligent choice of food is precluded."

Revolving Doors

Though business people involved in the mechanics of running a society at the governmental and/or political level is not a new phenomenon, the extent to which businesses are manipulating events in their favour at the expense of the public's health and safety is worrisome.

"You remember this revolving door business in the States?" asked Arpad.

Dr. Pusztai was referring to a handout given at a 2001 conference put on by the Council of Canadians, "Science and the Public Good." It was there that I first heard Drs. Pustzai and Bardocz speak. This document was a table (see Table 1 beginning on page 71) showing how people moved back and forth between government positions dedicated to food safety and corporate positions dedicated to advancing business interests. For example, Lord Sainsbury, CEO of a major UK food retailer, was Tony Blair's science minister and also a major shareholder and investor in GM companies. In another example, this time from the United States, Michael Taylor worked for a law firm "representing Monsanto when it applied for FDA approval for Posilac. He moved to FDA deputy commissioner when it was determined that GE foods were substantially equivalent, and hence, not subject to additional testing; he also wrote the FDA's rBST labeling guidelines." This table demonstrates an intriguing migration of people back and forth between corporate and governmental positions, which raises questions of objectivity and vested interests.

Similar job movements occur in Canada, as well. In 1999, a secret deal fast-tracked the approval of two new *Bt* varieties of genetically modified potatoes – without adequate field tests. "Internal memos show that John Doessetor, the senior advisor to the then Minister of Health Alan Rock, was kept up to date on these negotiations - a highly unusual situation since the Minister's office is not supposed to be directly involved in product reviews [.... Two years later], John Doessetor was hired by Monsanto to be the corporation's top lobbyist in Ottawa, 'responsible for the development and implementation of

Monsanto's government affairs strategies in Canada.'"[50] Many other examples exist, the source for which is footnoted.[51] A particular area of concern with these "revolving door" jobs is that often people hold positions on boards and committees where they are given the power to make complicated scientific judgments, on our behalf, just by the virtue of their business status.

"So," continued Arpad, "[many people in key positions] have no real knowledge to be able to weigh it up for themselves. They see all these huge dollar signs. 'This is going to be the future.' Little people like me will have to be put out of the way. I'm a 'bio-terrorist' because I'm working against the interests of the State, and now this is politics. I do understand why they do it, but that doesn't necessarily mean I agree with it or condone it. It's really terrible."

Sound Science?

"Now, whether they will prevail or not, I don't know," said Arpad. "I haven't got my crystal ball here. My role here, in this business, is less and less in the lab and more and more to write and speak ... So, until and unless [safety of GM foods] has been looked at, I'm of the skeptical viewpoint. The biotech industry's [usual response is] that GM is safe and that's a fact. Now, I think that – if it is said by a politician, I understand, but these guys are supposed to be so-called scientists. I think they are what I would call lackeys. Their views are paid for by the biotech industry, and if they cannot come up with factual information – they [just] say that the safety is a fact. Now, what I say is always a factual piece of information. In science, if you agree with it,

---

[50] Lucy Sharratt, "One Potato Two Potato Five Potato None," *EcoFarm & Garden*, Summer 2003 ‹cban.ca/Resources/Topics/Potato-Watch/One-Potato-Two-Potato-Five-Potato-None›.

[51] Lyle Stewart, "Good PR is Growing," *This Magazine*, May/June 2002 ‹www.healthcoalition.ca/goodprisgrowing.html›.

Table 1. Collegial relationship between government and the GE industry (from Borger, 1999; Ferrara, 1998; Independent-London, 8 March 99; Kingsnorth, 1998; Tokar, 1998)[52]

Individual	Moved From....	Moved To.....
Jack Watson	Chief of Staff in Carter White House	Monsanto Staff Lawyer
Michael Taylor	Lawyer for firm representing Monsanto when it applied for FDA approval for Posilac	FDA Deputy Commissioner when it was determined that GE foods were substantially equivalent, and hence, not subject to additional testing; he also wrote the FDA's rBST labelling guidelines
	Then back again from the FDA	to long range planning at Monsanto
Margaret Miller	Director of a Monsanto lab working on rBGH	Deputy Director of the FDA Office of New Animal Drugs

[52]E. Ann Clark, "Genetic Engineering in Field Crops: Ethics and Academia University of Guelph, " Saskatchewan Institute of Agrologists , Annual Meeting, April 1999 ‹www.plant.uoguelph.ca/research/homepage s/eclark/ethics.htm›.

<table>
<tr><td>Linda Fisher</td><td>Mapped pesticide policy in EPA under Bush</td><td>Monsanto's Vice President for Governmental Relations</td></tr>
<tr><td>L. Val Giddings</td><td>USDA/APHIS regulator</td><td>VP for Food and Agriculture at BIO (a Monsanto-backed organization)</td></tr>
<tr><th>Individual</th><th>Moved From ...</th><th>Moved To ...</th></tr>
<tr><td>Terrence Harvey</td><td>FDA</td><td>Monsanto</td></tr>
<tr><td>Suzanne Sechen</td><td>Graduate student at Cornell, working on several Monsanto-funded rBST studies, under a well known consultant to Monsanto</td><td>Primary reviewer for rBST in the FDA Office of New Animal Drugs 1988-1990</td></tr>
<tr><td>Keith Reding</td><td>USDA</td><td>Monsanto</td></tr>
<tr><td>Sally Van Wert</td><td>USDA</td><td>AgrEvo</td></tr>
<tr><td>John Gibbon</td><td colspan="2">Chair of Congressional Office of Technological Assessment, while at the same time a Monsanto consultant for more than a decade</td></tr>
<tr><td>Marcia Hale</td><td>Former Clinton assistant on intergovernmental relations</td><td>Monsanto coordinator of public relations and corporate strategy in UK</td></tr>
<tr><td>Micky Kantor</td><td>US Trade Rep and US Secretary of</td><td>Monsanto Board of Directors</td></tr>
</table>

	Commerce	
David Beier	Senior Director Government Affairs at Genentech, Inc.	Al Gore's Chief Domestic Policy Advisor
Fred Betz	Principal Scientist and regulatory specialist at the EPA, responsible for biopesticide risk assessment and biotech policy	Senior Scientist with environmental consulting firm assisting with GE registrations, and member of the newly appointed NRC GE review panel
John Podesta	Clinton White House Chief of Staff, whose lobbying firm represents pharmaceutical and chemical industry organizations, including Genentech	
Individual	**Moved From/To**	
Nick Weber	Researcher with the USDA (his supervisor is Margaret Miller), who passed confidential Codex documents to Monsanto, prior to the Codex meetings in February 1998 that approved milk from BST-treated cows	
Lord Sainsbury	CEO of Sainsbury's (major UK food retailer), Science Minister in the cabinet of Tony Blair, and is also a major shareholder and investor in GM companies	
Lord de Ramsey	Chair of the Environment Agency of the UK, who is being paid by Monsanto to test GMO crops on his land holdings	

fine. If you disagree with it, that's fine – providing you do some experimental work to prove that your viewpoint is right. Then eventually from all these conflicting things, we will come to some sort of understanding and a new dogma will be born. But this is not happening."

At the time of the scandal, Dr. Pusztai's research was the only known feeding study of GM potatoes. Since then, two other studies have come to light that seem to support the concern expressed by Pusztai. These studies also conclude that genetically modified potatoes are, so far, unsafe to eat.[53] However, ever the exacting scientist, Pusztai pointed out in a 2006 letter commenting on one of these studies that further corroboration of the study results was required[54] – something not likely to happen when GM products are such a hot commodity for industry. With that, and the fact that there is less and less funding without strings, it becomes very difficult for scientists to carry out full, accurate, and objective studies.

"You have to understand," said Arpad, "that all academia is very short on money and we are more or less being sold out to industry." With state budget cuts in research, universities and research institutes have to find the money somewhere, hence Arpad doing that interview on UK national television.

"There is absolutely a huge difference between state or publicly sponsored research and research by a single company," continued Pusztai. "The company is willing to put money into product development, but they are not going to spend money on basic science. There is also another thing; they want to keep the results. They don't want to share it with their potential competitors. This is an absolute total

---

[53] "GM potatoes unfit for human consumption," ‹www.ghorganics.com/GM%20Potatoes.html›

[54] "Pusztai responds to ACNFP over Ermakova," 19 Jan. 2006 ‹irina-ermakova.by.ru/eng/oth/otg32..html›

negation of scientific research. Scientific research is a communal effort where there is a free exchange, because you cannot do everything. It's impossible. So you are dependent on your colleagues producing results. You know the same things from different angles. When you are exchanging data, you get ideas on how to develop it further."

This is how things work in an ideal, scientific environment, "but if you put a total blockage in this process, then everyone is for him- or herself," Pusztai said. "So, it is total anti-science. I wish we had less research then to have this sort of research that is so highly focused that it will only go into the superficial areas of the whole business, and even then it's kept locked up. It's not a useful type of research. So, we might as well not do [it], because the other side of the story is that there is a tremendous amount of pressure on the scientists carrying out this highly focused, concentrated research. If you're producing something that is not to the liking of the sponsor, then you will be in real trouble, because they will just say that's not good enough for us. So, they'll stop and you'll be left high and dry, with nowhere to go. So, I think that is, in every sense, anti-science."

Genetic modification is an incredibly difficult scientific process to fully understand for lay people, but an analogy Dr. Pusztai gave makes clear his concerns about the level of science being used. According to him, gene manipulation, as it is now practiced, is akin to shooting in the dark and hoping you land in the right spot.

"This is my standpoint – that based on our experimental work and looking at other people's experiments, this what they call genetic modification technology, as it is done today, is not even a technology. At best it can be called a technique. Technology is something where you can predict what's going to happen. A technology builds bridges, builds houses, and in most instances with quite reasonable certainty you can say what sort of load that bridge, for example, will take, and it is designed in a predictable way. Now I challenge anyone to say whether a genetically modified plant is made in such a predictable, well-thought out way that it could be called a technology," said Arpad.

What Now?

"Now what is the way out of all this?" mused Arpad. "I think while they are using this so-called technology, they will get away with it for awhile, but basically it is unscientific. This genetic modification technique they are using is unpredictable, unsafe, and it will not produce anything that will withstand the judgment of time. Instead of this, we all ought to concentrate on different ways – I'm not against genetic modification. I'm against this particular one because it does not produce a scientifically sound product. So, the logical conclusion from this is we ought to have a look at whether there are other ways of doing this … but you see these companies want to sell their products. They're not interested in actually finding out."

Understanding this, people in Europe and the UK have made it quite clear to their governments that they do not want GMOs on their shelves and in their markets. Desi Boyle, a small store owner in Northern Ireland, was definitive on the subject. There's no way that he would ever get involved in genetically modified anything, he said. "First of all, the people don't know enough about it," Desi said, "and there's just nothing that would interest me personally, because to change something that doesn't need to be changed, doesn't seem right."

In general, there seems to be more awareness in Europe than in North America when it comes to GM issues. European citizens are more vocal about what they will and will not eat. However, while some foods destined for human consumption are still GM-free, the GM crops that are currently grown are being used mostly for animal feed. Because of the Mad Cow disease scare, Britain and Europe switched from animal-based feed to vegetable-based feed.

When Arpad was giving his fated interview on the BBC back in 1998, the first shipments of genetically modified soybeans were arriving in Great Britain to feed chickens, cows and pigs. Remember, though, that while GM soybean is certified "substantially equivalent," as are all other approved genetically modified plants, that does not mean it is safe to eat. Dr. Pusztai's study proved that rats fed GM potatoes developed pre-cancerous conditions in their internal organs, and therefore GM potatoes were unsafe for human consumption. If they were unsafe for both rats and us, then how safe could genetically modified plants of any variety possibly be for cows, pigs and chickens? What was their genetically modified soybean feed doing to *their* stomach linings? And, what effects might there be for humans to eat animals disfigured by the GM food that we feed them?

(Incidentally, the reason the GM corn Starlink had been banned for human consumption was that it was found to cause allergies in some people, but no one really knows if it's safe for us to eat animals fed with genetically modified products. Plus, if anyone were to have a reaction, it would be hard for doctors to trace since food labels don't always list GM ingredients. Sometimes that's because the manufacturer isn't required to list them, or they don't even know they are using genetically modified ingredients, as happened with GM potato ingredients in Japan, discussed earlier.)

In North America, except for a few relatively small organizations, there doesn't seem to be as much concern as there is in Europe. A GM potato variety modified by

Monsanto was approved by Health Canada "without any animal testing whatsoever."[55] In 2001, a mandatory labeling bill for GM products was defeated in Canada. "People who assume that there is actual testing ... will be amazed to learn that risk assessment of GM crops is largely heuristic or assumptions-based."[56] Even UK Rowett scientists Chesson and James stated that, in effect, it's better to be safe than sorry when it comes to health and food safety. Taking heed of these experts and taking action is the only sensible response to the problem of the genetic modification of our food supply.

"There will always be mavericks like me," said Arpad, "and I'm regarded a maverick, a whistleblower. I'm not a whistleblower, even though the effect was like that. I did say what I thought was right, and I had more right to say it than a lot of other people because I did actually do experimental work."

Not only is 2008 the International Year of the Potato as designated by the United Nations, it's the 10th anniversary of Dr. Pusztai's 150-second television interview. His confidence in scientific integrity may have been shaken by what happened to him, but maintaining a high level of scientific excellence, across the board, is extremely important to him. It is why he continues his work even in so-called retirement.

"As you can see, I'm 75," said Arpad. "I'm not particularly worried about it. I have so much work to do, it will take up my life."

...

Arpad's story is a chilling eye-opener about what has been done to the lowly spud and the lengths to which

---

[55] Smith, 232.

[56] Smith, 231.

governments and corporations go to pull the wool over everyone's eyes – including their own! Maybe it's time for us to get some teeth, just like the Quechua, who have refused to allow corporations, foreign or domestic, to take food out of their mouth.

Elizabeth Johnston

Raymond Loo at the market with his organic potato variety, Island Sunshine

Island Sunshine

(The first variety registered in Canada by a private breeder from PEI)

(première variété homologuée au Canada par une obtenteur indépendant de l'Î.-P.-É.)

Gerrit and Evert Loo, original breeders of Island Sunshine

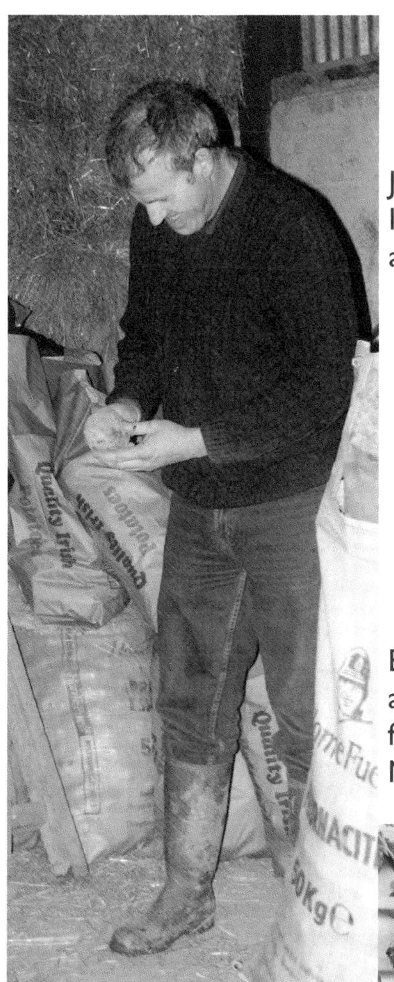

John Shepherd, Northern Ireland farmer, admiring his spuds.

Below, Stephen Bell, agronomist from Greenmount College in Northern Ireland

Drs. Arpad Pusztai and Susan Bardocz in Hungary.

View of Lake Balaton, from Dr. Pusztai's home in Hungary.

Árpad telling me about the *fogas* (pike fish) in Lake Balaton

Sampling Sandor's wine

At the PoPa. Alejandro, centre. Me on the left.

Mother Earth's breasts in the Sacred Valley

Quechua growing grid. Photo courtesy of IPBN.

Quechua children learning about their PoPa.
Photo courtesy of IPBN.

Barefoot technicians showing their spuds.
Note the one on the extreme right.

Eating *chuño* soup with little Anderson at left

Market shaman in Puno, Peru

A few of the potato varieties available at the market in Puno, on the shores of Lake Titicaca.

A few varieties at the CIP. Note the red "wedding ritual" ones, second row from left.

Ana, lab department head at CIP, in the cold vault that contains "Andean Treasure" – potato plants!

Sculpture on the shores of Lake Titicaca, represents the potato spirits and the agricultural seasons.

Potato martini,
enjoyed in Charlottetown, PEI.

Below, Uros altar on a floating island. Note the potatoes in thebowl on the right.

Chapter Four – The Heart of the Family

On the shores of Lake Titicaca, the highest navigable lake in the world, there are four sculptures representing the agricultural seasons. One in particular represents the potato's importance. "The light blue violet color," reads the description plaque at the bottom of the piece, "represents the color of the flowers and the ancestor spirits of the potato plant sowed in November. These plants bloom in February." Incorporated into the sculpture are two musical instruments, the charanga (a stringed instrument) and the pinkillo, (a whistle). "The charanga is an allegory to the dances in November and the pinkillo to the agricultural dances in February."

In some areas of Peru, music and dancing is still part of agriculture, including different phases of potato growing. One such tune is called the *pasacalle*, and it's played to accompany the workers to the potato fields. "Afterwards, during the actual fieldwork, members of the community continue the job of potato farming in close connection to the music. The music is changed in accordance with the tasks being undertaken and the effort of the laborers."[57]

The pale violet blue of the sculpture was part of a spectrum of blues between the deep lake waters and the high pale sky. I was at an altitude of 12,530 feet above sea level. The air was thin, making it hard to breathe for those not yet acclimated, and the bright sun bore down strongly, countering the chill experienced so high up. Through the holes of the sculpture that mimicked a whistle's finger holes, I could see cameos of the land behind in the distance. Dusty beige mountains with slogans cut into them. These carvings were a common sight in Peru. They were made by people

---

[57] Liner notes from *Traditional Music of Peru 2: The Mantaro Valley*, Smithsonian/Folkways, 1995.

who carved their political leanings into the sloping foreheads of the mountains.

From where I stood on the pier, the sculpture stood tall and straight, at least eight feet high. The sculpture was suspended over the lake by a small gangplank. Beneath the sculpture lay a crust of pungent green algae, so thick small birds walked on it. Further out on the lake itself, the Uros, Native Peruvians, lived on rafts they built from the reeds that grow so abundantly in the lake. Jacques Cousteau plunged into the lake once, coming face to face with its legendary giant frogs.

I came here because the town of Puno and the surrounding area is considered the cradle of the potato. "Researchers have pinpointed Peru's Lake Titicaca district, near the Bolivian border, as the place where the earthy tubers originated some 8,000 years ago."[58] All potatoes are thought to originate from this fertile place and then spread through the land by cultivated and wild means. "Inca myths relate that Viracocha, the Creator, caused the sun, moon and stars to emerge from Lake Titicaca. He also created agriculture when he sent his two sons to the human realm to study and classify the plants that grew there. They taught the people how to sow crops and how to use them so that they would never lack food."[59]

History was palpable here. There was a pulse in the air: the distant thunder of dances that opened and closed the growing season of the potato, echoes of a song heard in my childhood while I ate my grandmother's potatoes. I understood now. It was a song that told of a people's determination to make a life they could be proud of.

---

[58] Gary Lee, "Behold the potato," *AARP Seguna Juventud*, Aug/Sep 2007 ‹www.segundajuventud.org/english/travel/2007-AS/07AS_peru.html›.
[59] "Origins," Peruvian-Canadian Chamber of Commerce ‹www.perucanadacc.com/RECIPE_History.html›.

This was a different land than I was used to in Canada, but I could feel from where I stood, it was a place with teeth, a place where stories grew out of the soil, not just potatoes, and a place where human ingenuity flourished. From this fertile soil springs the story of the Quechua's Potato Park, an example of how a group of people working together managed to stop the walls of industrialized farming and genetic modification from closing in on them.

Meeting Alejandro

Early one quiet Saturday morning in September of 2004, I walked from my hotel in Cusco, Peru, to the headquarters of the Indigenous People's Biodiversity Network (IPBN), run by Alejandro Argumedo. He is a native Quechua, one group of Peru's indigenous peoples and has a B.Sc. in agriculture from McGill University in Montreal, Canada. Alejandro left Peru at the height of the Shining Path troubles. Many indigenous people lost their lives, even those who lived in remote areas of the country.

"The internal war in Peru was one big reason [that I left]," admitted Alejandro, "but also the lack of opportunities and the repression of dreamers, especially indigenous dreamers, we experience in Peru since the Spanish invaded the country."

Going to Canada meant Alejandro could realize his dreams, one of which was to help his people in Peru.

"I lived and worked in Canada (and also worked in Asia, the Pacific, Africa and the Americas) for about 15 years," Alejandro told me in an e-mail. "I was Executive Director of Cultural Survival Canada in Ottawa and have done some work with IDRC establishing an international programme to protect traditional knowledge. I am a member of the Board of Directors of RAFI and other international organizations. I am back in my native country since 1999 and working on the Potato Park idea since then."

I had come to see the Potato Park (PoPa) and learn more from Alejandro about how they had accomplished the seemingly impossible. "Quechua communities in the Pisac/Cusco area of Peru (an area characterized by rain-fed high altitude agriculture systems) have been working for several years to establish a 'Parque de la Papa' (Potato Park), a community-based, agrobiodiversity-focused conservation area. The initiative has brought together 8,000 villagers from six communities (Pisaq, Cusco, Saccaca, Cuyo Grande, Amaru, Paru-Paru, Pampallacta and Chawaytire), who have agreed to manage jointly their 8,661 hectares (about 33 square miles) of communal land for their collective benefit. Their aim is to conserve their landscape, livelihoods and way of life, and to revitalize their customary laws and institutions," read IPBN material.

What's in a Name?

For the ride out to the PoPa, five of us piled into the jeep. In addition to the driver and myself, there was Alejandro, his secretary, Sandra, and Chantaleesa, an American geography student studying the steppes farming techniques of the Quechua. As we pulled into the main playa in Cusco, we passed a huge church that had scaffolding erected in front of it, and a large drooping sign with the word 'vaccinations' in Spanish written across it. There was already a long line up into the church.

Alejandro mentioned the discrimination that still goes on today against the Native Peruvians. Officially, every citizen of Peru is entitled to health services, but the only language spoken in government offices is Spanish. Many Quechua do not know Spanish and, according to Alejandro, are therefore refused services since there are no representatives in government offices that speak any of the indigenous languages. So, many indigenous people go without services they are legally entitled to, such as immunizations, because of this discrimination. Also, Alejandro knew of instances where when the Native Peruvians would go to register their children's births, the officials would write down whatever name came to them since they could not understand Quechua. So, Quechua children were given astonishingly disrespectful names such as "Chair" or "Helicopter" on their birth certificates.

Historically, the Native population of Peru has had a difficult time ever since the Spanish invaded South America and used the potato to suppress them. In the 1500s, when the Conquistadors invaded and forced the Quechua men to work in the silver mines, the women tilled the fields, providing fresh food for the conquerors, but the Quechua were made to subsist on *chuño* (dried potato). Not only that, they were forced to pay for the *chuño* that Quechua women had made.

"The silver mines of Potosi," wrote historian Salaman, "discovered in 1545, were, of course, manned by native workmen, of whom, in the colonial period, untold thousands are said to have perished by reason of their ill-treatment in its deep and dust-laden galleries. These slave-workers were maintained almost exclusively on *chuño,* and bitter is the complaint raised by Cieza de Leon against the middlemen who swarmed out of Spain, bought *chuño* cheaply from the producer and, after selling it at a high price to the native workers, returned home with their ill-gotten fortunes. Potosi was no exception. Hans Sloane, after his return from the West Indies, informed the Fellows of the Royal Society that this method of 'subsisting' slave labour had been adopted in all the Spanish mines in Peru and elsewhere."[60]

I was beginning to see how the creation of the PoPa was even more remarkable than I had originally thought. Not only did they have to fight against foreign GM

---

[60] Salaman, 40.

corporate concerns, they also had to work within a system that did not value them as equal citizens, and in fact, treated them with less regard than the potato itself.

Re-education

Alejandro's education in Canada meant that he could be a bridge between the agricultural communities of his people and the technological world "out there" that sought to take control over the Quechua's very means of survival. In 1999, he returned to a community fractured by the disappearing culture, pressures of modernization, and the lure of work in the cities, far from their homes. There had even been some murders over border claims, something that was completely out of character for the Quechua, according to Alejandro.

Their worldview is based on one of sharing, similar to the ancient Irish *clachans* where land was not owned, but shared. For the Quechua, the idea of owning land was an idea imposed by the conquering culture. Several of the individual Quechua communities recognized that something drastic had to be done to block the tide of these destructive influences. The seed of an idea was born, and out of it eventually came the Potato Park. Creating a viable economic alternative was crucial if the Quechua were to stem the flow of their people to the city where they were absorbed by the mainstream, their culture and traditions diluted as if in an ocean. The PoPa would mean a rejuvenation of their culture, giving them back a sense of meaning and dignity that is so important for the human spirit.

Only six branches of the Quechua decided to be involved in the PoPa, but it was potentially over 8,000 more that were not reduced to sitting on hard city concrete hoping for the kindness of strangers. When I was in Lima, I stayed in the downtown core and near the hotel, there was a Quechua woman sitting on the sidewalk begging. Her bright full skirts pooled around her on the sidewalk. She stuck out in stark contrast to her environment and to those Quechua in the touristy parts of Cusco who posed for pictures in their bright costumes in exchange for money. They were poor, too, but unlike this rural woman displaced into the city, they were providing a service of sorts to tourists who wanted to take home an "authentic souvenir."

Alejandro underscored the importance of this difference when, on the way to the PoPa, we stopped to look down into the Sacred Valley. "The Sacred Valley of the Incas is actually the Vilcanota/Urubamba River valley. It is located about 10 miles north of Cusco, Peru, and extends northwest through Pisac and Ollantaytambo. This entire region, highlighted by Cusco, was the heart of the Inca civilization from the 14th to the 15th centuries."[61] Alejandro pointed out to us the way the terrain curved near the river in the shape of Mother Earth's breasts. Years ago, he said, farming was done only higher up on the mountain, and the lower land near the river was used just for rituals but now they farm there too, though they do work around the breast-shaped landmass to preserve the outline. The scene was breathtaking. It was a bright day with only a few clouds high above us. The air seemed to shimmer with the immense fertility of the area. The mountains spilled towards the river, pooling into distinct shapes that brought to mind the elongated breasts of a mother who had suckled more children than any one woman could. Farms extended along the banks of the river, and behind them were houses where the locals lived. The arrangement reminded me of the Irish *clachans* where the homes were built in a cluster and away from the shore.

While there, two little Quechua girls in traditional dress came up to us, one carrying a baby alpaca. They stood together as if we were to take their picture but none of us

---

[61] "Peru," Peru Travel Adventures ‹www.peru-travel-adventures.com/sacredvalley-cuzco.html›.

did. We took pictures of the landscape, instead. Alejandro chatted with the girls, until we finished. As we moved about on the gravel shoulder that had no guardrails protecting us from the sheer drop, I noticed that the girls stared at us with perplexed expressions.

Afterwards, as we drove higher up the mountains towards the PoPa, Alejandro said to us how once he and a friend had gone out to snap pictures of tourists "to see how they liked it." He explained that he had joked with the girls that they'd have to pay the visitors money if we took pictures of them. I suppose Alejandro was teaching the young girls a lesson that they might not understand until years later. It could be seen as a seed of dignity that he was planting, and it was also a subtle way of connecting for us the thousands of years of oppression that the Quechua have already endured to what continues to happen today. In a way, taking a picture of the Quechua and not compensating them for what we've taken from them without their permission is not too far removed from the unfair treatment they've received for centuries. That minor, fleeting moment had me thinking about the importance of not perpetuating economic disparity, and as well, that the PoPa was a way for the Quechua to move beyond that disparity. Through it, they could create their own wealth instead of relying on others to provide it or take it away. By returning to the potato, they were providing themselves with cultural and food security, something that would disappear with genetic modification.

Repatriation

To create that security, one of the first things the Quechua needed to do was get their potatoes back from the International Potato Centre. Located in Lima, this institution is a product of the Green Revolution, a movement in the 1950s that advocated better agriculture through chemicals and science. "The International Potato Center (known by its Spanish acronym, CIP) seeks to reduce poverty and achieve food security on a sustained basis in developing countries through scientific research and related activities on potato, sweet potato, other root and tuber crops, and on the improved management of natural resources in the Andes and other mountain areas."[62] CIP is the largest repository of potato genetic material in the world, material that is to be held in trust for all humanity. Because of its importance in the preservation of the potato, it is funded by "58 governments, private foundations and international and regional organizations."[63]

In their cold vault that is maintained at a temperature of between 6 and 8 degrees, there are over 10,000 test tubes of tiny potato plants. "We like to call this Andean Treasure," said Ana, the lab department head. Some of the plant samples come in with viruses or disease, and they have a special process to make them healthy, which includes putting them into the cold vault. This slows down their growth but also reduces the population and density of the virus in the plant. After a certain amount of time, technicians take a tiny sample from the top of the plant because that will have the least amount of or no disease, and then they re-grow a healthy plant from that sample part. Theoretically, the new plant is then ready to be farmed again.

However, that wasn't happening, according to Alejandro. He explained how decades ago, CIP officials came into the area and took hundreds of potatoes to their lab to study them but didn't give the new and improved strains back or share any of that knowledge until Alejandro came along and began the process of repatriation.

---

[62] "What is CIP?" ‹www.cipotato.org/cip/about.asp›.
[63] CIP.

Alejandro worked to bring those potatoes back because he believed the only way to bring back the culture of the Quechua was to bring back the spud. "Genetic diversity and cultural diversity are closely linked. When the potatoes came back, the culture came back,"[64] he said. Finally, in December of 2004, the PoPa and CIP signed a potato repatriation agreement. Only then could the slow process of actually getting all the potatoes back again begin.

Since the Quechua think of the potato as a member of the family, each recovered potato must have been celebrated with joy. It certainly was with pride and playful gleams in their eyes that the Quechua showed me the cold room where a few dozen of the already repatriated potatoes were kept. This was a small room in the PoPa headquarters with racks against two walls. One wall held the potatoes that have been repatriated, the other, potatoes that never left.

Several of the barefoot technicians each picked up a potato to show me and Alejandro translated for them. One of the potatoes was a red, bumpy one, sort of like a bulbous rosette. "The gnarly and dark *huaahat* – roughly translated from Quechua as 'that which makes the daughter-in-law weep' – is presented to the bride-to-be at the wedding ceremony; her peeling skills become a kind of marital litmus test. The *ohasito,* small and bright purple, is said to act as a natural Prozac and is used to combat depression."[65] One gentleman held up a potato that was particularly long and thin. We all laughed heartily. I didn't need any translations for that, but it brought to earthy life Salaman's emphasis on the Quechua's association between sexuality and the spud.

---

[64] "GM potatoes expelled from Andes," *Nature News* ,16 July 2007 ‹ise.arts.ubc.ca/about_us/members_argumedo.php›.
[65] Lee, "Behold the potato."

Ground Level Mobilization

While Alejandro was flying back and forth over the Andes Mountains between Cusco and Lima negotiating the return of the potatoes, people at the PoPa were kept busy, too, rebuilding their almost lost culture. Over the years, many of the young had either gone to the cities to make their fortunes or stayed in the country but adopted a mainstream life, rejecting farming for 9 to 5 jobs that were removed from their traditions. As a result, much of the old knowledge, accumulated through countless generations, was disappearing. As the elders were dying off, so were their stories and wisdom. So, part of the work necessary to create the PoPa was to organize people into different groups. There were the barefoot technicians, experts in cultivation and varieties. These were the people who showed me their spuds in the cold room. There were the storytellers, people in charge of remembering the culture's stories and passing them on, as well as herbalists, videographers, and others.

People from these different groups worked in various rooms at the PoPa headquarters. In one room, there was a computer where some people were busy compiling their database of the varieties they grew along with pictures and growing information. The data also included information taken from their matrix which is a grid with symbols on it that instruct when to plant, and how to grow and protect each variety. It also lists information about things like beneficials – plants, like oca, that are good to grow alongside the potatoes to protect them from slugs or other pests. As well, information about their medicines went into the database for safekeeping. "A database of traditional medicinal knowledge is also being established to protect against biopiracy, and a network of barefoot technicians, who are elected by their communities because of their expertise in traditional knowledge, are developing a dynamic process of horizontal learning and knowledge exchange. These barefoot technicians have been supporting other communities by providing information, supporting exchange of experiences and cross-community visits, offering participatory planning and evaluation methods at the community level, organizing training courses and helping to advocate for the needs, visions and rights of indigenous peoples and their knowledge systems."[66]

Intellectual Property

Everyone has a role to play in the success of the PoPa, Alejandro emphasized, and the women who were involved in developing Quechua video registries had a particularly important role in ensuring that what the PoPa had begun will continue. The video documents they produce are of value for their agricultural, historical, and cultural aspects, but are also equally important for their political and legal value. They are meant to protect their potato varieties as intellectual property against poaching from those who would like to patent life forms. Just how important this issue is to the cultural survival of the Quechua was underscored by an incident during the earlier introductions.

Once everyone was assembled in a wide circle in the kitchen, we introduced ourselves with Alejandro translating back and forth. When Chantaleesa, the geography student from California introduced herself, the Quechua laughed because, they said, her name meant 'angry singer' in Quechua. They also joked that they had better watch out for the American because she might take something back to be patented. We quickly passed onto other topics, but the worry beneath their humour was obvious: the future of the PoPa was far from secure in the face of foreign corporate concerns.

[66] "Peru: Protecting Rights, conserving diversity," Indigenous Peoples Biodiversity Network, August 2006.

Those concerns have their roots in America's foundational mythology – the frontier. When the first settlers landed on the North American continent, it was wild and untamed. People moved west, cutting down trees, and sometimes Natives, as they moved west, until they met the ocean on the other side and then there were no more frontiers to conquer. Now the frontier has gone to another level, an interior level, the level of food and DNA and patents. The people chasing after these new frontiers, seeking to conquer and name – and more importantly own – everything that crosses their path is another way of trying to control culture and identity. Now that myth seems to have informed the rise of biotech corporations around the world.

No sooner did the Quechua find a way to extricate themselves from the long-term effects of Spanish colonization than another type of conquistador is charging over the distant mountains – biotech and pharmaceutical companies like the American giant, Monsanto. It's a clash of cultures that requires fighting fire with fire. In other words, to protect the potato and those it feeds, contracts and agreements need to be brokered that have real teeth.

Owning Life

One approach is for farmers to use intellectual property law to protect their rights. "The term 'intellectual property' refers to a group of laws such as patents, plant breeders' rights, copyright, trademarks and trade secrets that are intended to protect inventors and artists from losing control over their intellectual creations – their ideas. Intellectual property has become a powerful tool to enhance corporate monopoly and consolidate market power. Monopoly control over plants, animals and other life forms jeopardizes world food security, undermines conservation and use of biological diversity, and threatens to increase the economic insecurity of farming communities."[67]

Philosophically opposed to the idea of intellectual property, Raymond felt there was no choice, given that there were no other alternatives available to protect Island Sunshine from being co-opted by a company like Monsanto. The Loo family consciously chose to go for breeders' rights instead of a patent because they wanted to make sure that others could have easy and inexpensive access to what the Loos had discovered. Following in his father's footsteps, Raymond is more interested in doing something to help the organic potato industry and the world rather than make a lot of money, which is the *raison d'être* for corporations. Additionally, having a patent meant owning life, and that is something the Loos are also diametrically opposed to.

"I don't like the idea of owning something I don't really feel I own," said Raymond. "I don't feel I really own this earth, our farm. I don't own it. I have a privilege to be able to farm it. But, if you think of it, how can you own it? You can't. You can buy a car. You can own that. But you can't really own life, in my opinion."

Still, Raymond remains ambivalent about having had to work within the existing system.

"On the one hand, I can see where people should pay to help us continue the work, but on the other hand, I just don't like the idea of breeders' rights," said Raymond.

The Quechua don't either. "Intellectual property regimes and international trade agreements legally condone patents and activities that are predatory on the indigenous knowledge or sovereign genetic resources of other people,"[68] and so those involved with the PoPa refuse to participate.

---

[67] "Intellectual property rights and patents," ETC Group ‹www.etcgroup .org/en/issues/intellectual_property_patents.html›.

[68] Hope Shand, ETC Group, ‹www.etcgroup.org/en/materials/publication

"These indigenous people are against patents," Argumedo explained. "They represent a model of property that does not fit into their worldview. Indigenous people are used to exchanging and sharing information in open ways."[69]

Though the Quechua worldview does not accept patenting, they recognize that they need to protect their cultural heritage. Their video registries, computer database, as well as the agreement between the PoPa and CIP are a start.

Alejandro reflected on the importance of what the Quechua had achieved with this agreement. "This means a legal agreement that no one else can claim intellectual property rights over their knowledge," said Alejandro. "The implications can be far-reaching [because it] is the first legal sign of the restoration of rights that indigenous people once had."[70]

That the agreement is with CIP, the world gene bank of potatoes, a respected organization worldwide, goes a long way to legitimize indigenous rights and knowledge globally. This is especially so for the many poorer countries that are battling mega-corporations over resources that have been patented right from underneath them. The term for this is called biopiracy.

Biopiracy Cases

Biopiracy "describes a new form of 'colonial pillaging' in which western corporations reap profits by taking out patents on indigenous plants, food, local knowledge, human tissues and drugs from developing countries and turning

---

s.html?pub_id=690›.

[69] Sanjay Suri, "Potato Capital of the World Offers Up New Recipe," *IPS News*, 18 Jan. 2005 ‹www.grain.org/bio-ipr/?id=429›.

[70] Suri, "Potato Capital."

them into lucrative products."[71] There are many cases of biopiracy worldwide. Here are just a few examples.

In the 1990s, Biogaia, a Swedish biotech company, "patented a microorganism extracted from the breast of a Peruvian woman and the fungus was commercialized in yogurt and other products being sold in Scandinavia."[72]

In 1999, American citizen Larry Proctor secured a patent on the enola bean, better known as the Mexican jumping bean. "After securing his monopoly patent, Proctor accused Mexican farmers of infringing the patent (US patent number 5,894,079) by selling yellow beans in the US. As a result, shipments of yellow beans from Mexico were stopped at the US/Mexican border, and Mexican farmers lost lucrative markets. In 2001, Proctor filed lawsuits against 16 small bean seed companies and farmers in the US, again charging patent infringement."[73] The patent was challenged and investigators found that Proctor's enola was not a new or novel invention (the requirement for a patent) and in fact it was the exact same as an existing bean. Procter eventually confirmed that he had brought back a bag of enola seed from Mexico to the United States but claims to have subsequently produced a seed with a distinctly yellow colour.

In April 2008, Proctor's patent was rescinded, but under international intellectual property laws, no restitution is required of the offender to compensate the farmers who were prevented from pursuing their livelihood from 2001 to 2008. The cost of taking the case to court is also a major

---

[71] "Biopiracy: The New Colonialism," Conglomco Media Network, 16 May 08 ‹cong lomco.org/serendipity/index.php?/archives/3340-Biopiracy,-the-new-colonialism.html›.
[72] Alejandro Argumedo, "Yacon Come Home?" ETC Group, 10 Aug. 2001 ‹www.etcgroup.org/en/materials/publications.html?pub_id=254›.
[73] Hope Shand, "Hollow Victory: Enola Bean Patent Smashed At Last (Maybe)," ETC Group press release, 29 Apr. 2008 ‹www.etcgroup.org /en/materials/publications.html?pub_id=683›.

deterrent for most small farmers, unless they have the support of other groups. This situation is likely to increase as the incidence of expensive lawsuits stemming from patent infringements is on the rise.

In terms of lawsuits instigated by Monsanto in the United States, a 1995 study found that to date, "Monsanto has filed 90 lawsuits against American farmers in 25 states that involve 147 farmers and 39 small businesses or farm companies." Given that in 2008 (13 years after this study was released), the average North American farmer makes less than $50,000 per year, it's not likely they'd have a lot of disposal income to defend themselves in court against a corporate giant like Monsanto. However, "Monsanto has set aside an annual budget of $10 million and a staff of 75 devoted solely to investigating and prosecuting farmers."[74]

Two farmers that have gone head-to-head with Monsanto told their story in the documentary, *The World According to Monsanto*. One of them, Troy Roush, grows primarily conventional (non-GM) soybean. Troy, however, did agree to grow a small area of GM soybean on contract. One day a Monsanto private investigator showed up and said that Monsanto suspected them of growing GM seed illegally. Despite the proof Troy had to the contrary, Monsanto took him, his two brothers and his father to court. Eventually, he settled out of court. Regardless of the fact that Roush insisted they were innocent, the prospect of losing the family farm and his children's legacy made them get out of the court battle before everything was lost.

"After two and a half years of this, the family was just destroyed," said Troy. "The stress involved in this ... They're in essence threatening five generations of work, and

---

[74] Craig Culp, "Monsanto Assault on Farmers Detailed in New Report," press release, Organic Consumers Association, 13 Jan. 2005 ‹www.organicconsumers.org/monsanto/assault011405.cfm›.

if they were to prevail in something like this, it's all gone. They'd take it all away."

Troy feels that Monsanto is trying to take over the seed market completely. Joseph Mendelson of Washington's Center for Food Safety (CFS) agrees.

"Monsanto would like nothing more than to be the sole source for staple crop seeds in this country and around the world," said Mendelson, legal director for CFS. "And it will aggressively overturn centuries-old farming practices and drive its own clients out of business through lawsuits to achieve this goal."

Rodney Nelson from North Dakota is another farmer who was sued by Monsanto. "Monsanto is taking advantage of farmers with their marketing and their threats and lawsuits," he said. "It's hard enough to farm as it is. You don't need a big seed supplier trying to trip you up and chase you down with lawyers."[75]

The CFS reported that "the largest recorded judgment found thus far in favor of Monsanto as a result of a farmer lawsuit is $3,052,800.00. Total recorded judgments granted to Monsanto for lawsuits amount to $15,253,602.82. Farmers have paid a mean of $412,259.54 for cases with recorded judgments. Many farmers have to pay additional court and attorney fees and are sometimes even forced to pay the costs Monsanto incurs while investigating them."[76]

In Canada, the most famous occurrence of a Monsanto lawsuit against a farmer involves Prairie wheat farmer, Percy Schmeiser. It began in 1998 when Monsanto sent a letter to Schmeiser saying that investigators had found GM wheat in his fields and demanded payment. When Percy refused, saying that the wind must have blown the

---

[75] Culp, "Monsanto Assault."
[76] Culp, "Monsanto Assault."

seed onto his fields, Monsanto took the farmer to court where the corporation won its case.

"I was really alarmed at the fact that it said in the decision that it doesn't matter how it gets into a farmer's field – whether it blows in or cross-pollinates, floods, comes in on farm machinery – it doesn't belong to the farmer. It belongs to Monsanto,"[77] Schmeiser said.

By this time, Percy wasn't sure he would appeal since the Schmeisers had used up their entire $200,000 retirement fund. Appealing the decision and losing would mean he'd lose his farm to bankruptcy. Still, he said that he was prepared to fight on. Then he broke down saying that he had to consider the impact on his wife, who had been ill.

In 2008, however, Percy won his case, with a lot of monetary and moral support along the way from individuals and organizations. The press release posted on his Web site sums up the victory decision. "In an out of court settlement finalized on March 19, 2008, Percy Schmeiser [now 77 years old] has settled his lawsuit with Monsanto. Monsanto has agreed to pay all the clean-up costs of the Roundup Ready canola that contaminated Schmeiser's fields."

For Monsanto, however, this must just be the cost of doing business. It was during the span of this one case, out of close to 100, that Monsanto achieved the status of the number one seed company in the world in 2004. This was about halfway through the 10-year ordeal that took almost everything away from Percy Schmeiser – who had done nothing wrong in the first place.

Pleased with the outcome, though, Percy described for a newspaper reporter the scope of the court's ruling.

"Also part of the agreement was that there was no gag-order on the settlement and that Monsanto could be sued again if further contamination occurred. Schmeiser believes this precedent setting agreement ensures that

[77] ‹www.percyschmeiser.com›.

farmers will be entitled to reimbursement when their fields become contaminated with unwanted Roundup Ready canola or any other unwanted GMO plants."[78]

You might think that such a decision would have ripple effects around the world, and that countries would automatically fix the laws so that it won't be "*when* their fields become contaminated." But that has not been the case. The wins that the little guys like Percy achieve are few and far between and may not be enough to save seed, farmers and consumers alike from the deleterious effects of genetic modification.

Even scientists from the Rowett Institute have their misgivings about the ability of individuals or even countries to decide their own fate in the face of the proliferation of genetic modification in the current economic climate. Dr. Chesson and Prof. James wrote in their report to the House of Lords of the difficulties the UK may face if they banned certain GM products from America. "If the EU concludes that a biotechnological innovation is unacceptable on safety or other, e.g. environmental, grounds, this may clash with current US, Japanese or Chinese assessments. The US government, for example, may then claim inappropriate trade discrimination and appeal to the WTO."[79]

Chesson and James cited a case in which India was taken to Free Trade court by the Americans. "Public Health has traditionally been of little import in WTO terms when there is a trade dispute. Thus when India tried to ban US cigarette imports, the WTO forced India to give access to US tobacco companies and use equivalent measures against cigarette use for both US and Indian products. Despite the huge financial muscle of US tobacco firms, the only measures available to India were to ban all cigarette

---

[78] ‹www.percyschmeiser.com›.

[79] James, memo ‹www.publications.parliament.uk/pa/cm199899/cmselect/cmsctech/286/9030810.htm›.

advertising and discourage cigarette use through a variety of classic measures such as health education and smoking restriction in some public places."

A similar thing happened in Canada when it tried to ban MMT, a gasoline additive that was found to cause neurological damage. "In the age of Free Trade however, prudence, logic and the public interest come a distant second to the rights of business. MMT's manufacturer, the Ethyl Corporation of Richmond, Virginia announced that it was suing the Government of Canada, claiming $251 million in damages."[80] In the end, Canada was forced to continue buying the product, pay the manufacturer $19 million, and issue a public statement that the additive posed no risk, after all.

"Under the rules of Free Trade," wrote Tom Fuller of the Alberta Federation of Labour, "the rights of an American corporation to make a profit outweigh the obligations of the government of Canada to take prudent precautions in the public interest."

Though there are people who are actively battling this illogical and unfair practice, it comes down to money. As was seen in the case of Paraguay and the many lawsuits against farmers, biotech companies have vast quantities of money they can use to push their products into communities whether those communities and/or individuals want them or not. And with the WTO in their corner, it looks like we might be forced to swallow just about anything.

Terminator Tater

Despite the bleak picture, Alejandro presses on. However, now the fight to save the Quechua way of life has reached

[80] Tom Fuller, "Government Humiliation in Free Trade Lawsuit," Alberta Federation of Labour ‹www.telusplanet.net/public/afl/LabourNews/sept98-9.html›.

new heights with the spectre of Terminator technology in potatoes. Terminator technology refers to genetic modifications to seeds that force plants to die after they produce their fruits only one time. Any seeds collected and replanted from a Terminator plant will not grow. In terms of potatoes, the biotech company Syngenta received a patent for its Terminator potato technology in 2004, US Patent 6700039. This patent "describes a genetic-modification process that could be used to stop potatoes from sprouting unless a chemical is applied."[81]

Alejandro, along with other members of the Indigenous Coalition Against Biopiracy, launched an international campaign against Syngenta's Terminator Potato.

"We feel greatly disrespected by corporations that make a single genetic alteration to a plant and then claim private ownership when these plants are the result of thousands of years of careful breeding by indigenous people," said Argumedo.

Even though Peru has recently passed a law banning any kind of genetically modified potatoes (whether it be the growth, sale or transport of them), there is the fear that illegally planted potatoes will cross pollinate with natural varieties and irrevocably damage the plant and the Quechua culture. Although potatoes don't breed as easily through pollination as other plants, it does happen, and "cross-pollination seems to be much greater when the GM and non-GM varieties are different and when the main pollinator is the pollen beetle, which travels far."[82] One particular

---

[81] "'Insulted' Andean farmers pick GM potato fight with multinational Syngenta," press release, International Institute for Environment and Development, 12 Jan. 2007 ‹www.grain.org/bio-ipr/?id=500›.
[82] "GM potatoes - summary of the risks," policy document, Soil Association ‹www.soilassociation.org/web/sa/saweb.nsf/librarytitles/22E7A.HTMl›.

study found that organic potatoes grown one kilometre away from GM potatoes had a cross-pollination level of 31 percent.[83] So, the threat to potato and Quechua sovereignty is profoundly real. In 2006, the Coalition sent a letter to Syngenta asking to discuss the issue, but no talks have been initiated.

"We received an insulting letter in reply," said Alejandro Argumedo. "Syngenta disregards our culture, values and our right to use the tubers of a resource that our peoples have nurtured for millennia. Introducing 'terminator technology' potatoes could create major problems for farmers in the Andes."[84]

As part of the campaign, the Coalition called on organizations and countries around the world to make their voices heard. In response, the secretary for the World Council of Churches issued a statement condemning terminator technology. "Preventing farmers from re-planting saved seed will increase economic injustice all over the world," said Samuel Kobia, council secretary, "and add to the burdens of those already living in hardship."[85]

Though there has been a worldwide moratorium on Terminator technology since 2000, corporations are still developing the technology, the laws of countries are still allowing this to go on, and patent offices are still issuing patents for this technology. "Syngenta has been granted terminator potato patents in Australia and Russia and has applied for similar patents in Europe, Brazil, Canada, China, Egypt and Poland."[86]

---

[83] Soil Association.

[84] "Insulted Andean farmers."

[85] "WCC Demands Action to Stop Terminator Seeds," press release, 15 May 2006 ‹www.etcgroup.org/en/materials/publications.html?pub_id=12›.

[86] "GE Potato Threat to South Africa," press release, Canadian Biotechnology Action Network, 10 July 2008 ‹ http://cban.ca/Resources/Topics/Potato-Watch›.

Eating Potatoes

When I visited the PoPa in 2005, I tasted the fruits of the Quechua's efforts in the PoPa. Alejandro ushered us into the kitchen where many Quechua were already seated, talking animatedly amongst themselves. I joined the others at a table just as tin bowls of steaming potatoes appeared in front of us. Alejandro dug in first. Then we all followed suit, picking up a warm potato with our hands and eating it. They were small potatoes, good for two or three bites. I ate mine with the skin on, biting right into it. It was a moist, earthy-tasting spud, and when I looked down at the other half in my hand, I saw the remains of a larva. No wonder it was earthy tasting! I put that potato down on the table, looking around, embarrassed. No one else seemed to have a larva in theirs, or be paying attention to me, thank goodness.

Looking back on that moment, though, I'd much rather come across the odd larva in my potato than have to eat potatoes that have been genetically altered to commit suicide after one harvest. Better that than to eat spuds that have insecticides and chemicals injected into their DNA, or foreign genes spliced into them that might cause cancer in my stomach lining. I'd rather eat a bug or two than to participate in the destruction of cultures and people's sense of dignity.

Rooted in the Future

After I put the potato with the larva down on the table, I noticed that not only did Alejandro peel his first, he also broke off a piece of potato with his fingers and put that morsel into his mouth. So, I followed suit, and did not find any other larva in the six or so potatoes I ate. As I dipped my potato morsels into a bowl of spicy peanut sauce called *huancaina* made from onion, pepper, alpaca cheese and milk, I noticed a little boy shyly clinging to his mother's skirt. I asked Alejandro what the boy's name was.

"Anderson," he said. "Soccer is very popular here, and so many children are named after a favourite player."

I ate the next course, *chuño* soup – made from that ancient dried potato which allowed the silver mine slaves to survive. Five hundred years later, here I was, a woman from industrialized Canada, taking into her mouth a piece of Peruvian history.

The potato in its fresh or dried form was, and is, of immense importance to the Peruvians. It provided them with sustenance and was an integral part of their cultural and spiritual lives. Despite the way *chuño* was used against them by the Spanish during the 1500s, it remains a staple that, like the spuds repatriated from CIP, the Quechua are reclaiming as part of their culture. The use of the word "repatriation" to describe this process shows just how much the Quechua are reacting against the conquest and subjugation of their culture and themselves. It is a noble word for a noble vegetable.

Even in dried form, this vegetable has nourished millions for thousands of years. I thought of how it had been doing that for the whole world since the Europeans brought it back with them to Europe. At first, it was looked on with suspicion but eventually embraced in France. During the Black Plague, the king was faced with starving masses who believed that the potato had caused the Plague. A member of the court, Louis Auguste Parmentier, returned from a Russian prison after seven years' incarceration and surprised everyone with how well he looked, considering what he'd been through. When he told the King that was because he ate almost nothing else but potatoes for seven years, an idea was born to solve the hunger ravaging the King's subjects. The King had his royal rose beds dug up and replanted with potatoes. Then he changed the tuber's name from *patate* to *pomme de terre*, and made a big show of posting many guards around this new garden. When the potatoes were ready to be lifted, the King ordered the posted sentries to pretend to look the other way when peasants came to steal the tubers in the middle of the night. The idea was that the peasants would surely steal the spuds since, "If it was good enough for the King to eat, it must be good enough for me to eat." The ruse worked and potatoes have never looked back since. (In fact, today they are the fourth most important crop worldwide and are even replacing some rice paddies in China. As well, there is a statue of Louis Auguste in the Paris metro, Parmentier Station. Louis stands with a basket of potatoes on one arm and a potato in an outstretched hand.)

In a very real way, the Quechua haven't looked back either. Their headquarters had all the accoutrements of modernity: computers, Internet, television, fridges and stoves. On the way to the PoPa, Alejandro had pointed out the foreign architecture – examples of homes built in the style the Spanish conquerors brought with them versus the adobe homes that the indigenous people built. Alejandro mused, "What would any young culture develop into, if they were given the chance? What type of modern architecture would arise organically out of the indigenous cultures if their natural growth had not been stunted by their oppressors, and in many cases even replaced?" That's a question the Quechua are in the process of answering as they work on the PoPa, blending history and modernity as they see fit.

As Raymond said when he talked about organic farming, "We're not going backwards. We're going forward with knowledge from the past." Irish historian Kevin Whelan talked about this way of combining the past and the present, too. Whelan distinguishes between nostalgia and radical memory. Nostalgia is looking back with regret and inertia – a wishing to go back but knowing it's impossible. Radical memory is bringing back from the past the lessons learned and integrating them into the present in order to create a better future. It is active, not passive like nostalgia. What Raymond in PEI and the Quechua in Cusco are doing is looking back for their values to root their present and build their future on.

Apus

Looking around us outside the PoPa headquarters, Alejandro pointed out the community's landmarks. From where we stood, we could see the villages below and the various fields, most of them for potatoes. Then Alejandro talked about the mountains.

To the Quechua, these physical landmasses are the mountain gods, their *apus*. Each of the six communities involved in the PoPa is associated with one particular *apu*. Anyone in the community can talk with their *apu*, but there is one person in each tribe that has a more intimate relationship with the *apu*, and that is the shaman.

When the six communities decided to work together towards the goal of the PoPa, their *apus* also joined forces. Within the larger context of the park, each *apu* works with the others to achieve their collective goals. For example, one *apu* might be in charge of making sure there is enough rain, another that there is enough sun. If an *apu* isn't doing his or her designated job (e.g. preventing drought) they can be jailed until they behave. The jail is located next to that year's controlling *apu*, which changes every year by election.

The relationship between the *apus* and the community is one of reciprocity. The gods can ask things of the community just the way the community can ask things of the *apus*. This equality is expressed in the Andean Cross where the spiritual realm is reflected in the earthly realm and vice versa. Interestingly, one exterior wall of the PoPa was a giant Andean Cross. Their spirituality was a literal and figurative support for their enterprise. The location of the

PoPa office was on a plain specifically chosen because it was equidistant from each of the *apus*, thus again underscoring the communal nature of the park.

The PoPa demonstrates that a community is not something nostalgic that would be nice to go back to. A sense of community and belonging is vital for an individual's sense of identity, whether in rural Peru or downtown Montreal. The trend set in motion from the Industrial Age is one where economics dictates how we live rather than people determining how economics will improve their lives. It isn't what's best for people that drives our choice of products; it's what's easiest and most profitable for the manufacturer. It's a case of us fitting into technology (in the broadest sense of the word) instead of technology fitting into our needs. And when you replace community values with economic ones, you end up with corporations using people to further their profit margins. The Quechua of the PoPa, on the other hand, are actively creating a place where the economy, like the *apus*, works for the good of the community in a reciprocal fashion. What the Quechua have achieved could be a possible model for other communities around the world. At the very least, it's proof that great things can be accomplished when people work together *for the common good.*

Chapter Five – Buried Treasure

When I was a child playing on my grandparents' property, I discovered wild flowers that I later found out were Moccasin Flower, also known as Lady's Slipper. What a delight that was for the dreamer in me to imagine forest faeries wearing these delicate orchids on their feet. I came across the flowers while my grandfather, Toni, worked nearby, spreading out the high piles of sand recently delivered. Toni wanted to sop up the swampy bits of land so he could open a campground. My grandparents did just that and eventually made a success out of the business. At the time, I remember tugging Toni by the sleeve to come see the flowers. Very seriously I made him promise not to cover them over. Humouring me, he obliged, but the next day when I came by to see if the faeries were cavorting in their slippers, the orchids were gone, buried beneath the sand. I was devastated and felt so betrayed by my grandfather. Those plants were important to me only because they looked lovely and set my imagination free. At that young age, I had no idea of biodiversity; it just seemed to me that there ought to be room for both business and beauty.

In the course of this project, I learned that the Lady Slipper is the provincial flower of Prince Edward Island. How serendipitous that the flowers I felt so protective of in my childhood would be connected to the island I would visit later in life for research on this book. My childhood experience with the Lady Slipper was a seed that grew into the journey I would take to the Potato Island of Canada and beyond.

Vandana Shiva, physicist and agricultural activist from India, wrote that "the seed ... is not merely the source of plants and food; it is the storage place of culture and history." It is also the storage place for creativity, new projects, and stories that give meaning to our lives.

Personal Stories

Culture is created from traditions passed down and through memory. The childhood stories we take with us into adulthood and pass on to our children or others through art, music, dance, et cetera, shape who we are and how we communicate with the world. That's why the Quechua have a group in charge of storytelling.

Desi Boyle, the owner of a small grocery store in Antrim, Northern Ireland, has his own potato story. He worked his way through university by selling potatoes. Desi took a little van, went to the local farmer and bought several bags of half-hundred weights, (56-pound bags). Then he bought a package of brown paper bags and a scale. "A lot of people would buy a weight," recalled Desi. "We just had a brick, broke it in half and it would be about five pounds, put that on the scale, fill up the other side with potatoes and do it again until you got 10 or 15 pounds for the customer."

"The scale Desi used was an old-fashioned brass scoop," explained my research assistant Fred, "and if you take everything off, it would be level. If you put a five-pound weight on one side, fill up the hopper until it balanced, and the needle went to zero. You pour those potatoes into the bag and do it again."

"Incidentally," recalled Fred, "that's what our neighbourhood potato farmer used to weigh the potatoes with. He used to come around our district in Ardoyne, Belfast, with his little van selling his produce. Sometimes, he wouldn't have any bags, so my mum said go get something to put the potatoes in. I ran into the house worried that Neisan the farmer would leave before I could get back out there. So, I grabbed the first thing at hand – my father's cap." Afterwards, he put the cap back on the hook, but forgot to clean it first. When his father put his cap on the next morning to go to work, bits of soil fell out all over his father's face and clean work shirt. The young Freddie was soundly punished for making a dirty mess of his Dad's cap, but it's a memory that has stayed with him.

In these stories are hands moving across these years, hands that helped each other or provided opportunities to learn about the world and how to succeed in it – that farmer growing his spuds, then selling them to Desi who sold them to his city customers, which enabled him to pay for his university tuition. Now he runs his own grocery store. There are also the hands of a little boy who brought potatoes to his mum so she could wash, peel and cook them for the family dinner.

There's an ordinary beauty in that, the kind that Vincent Van Gogh celebrated in his series of paintings called *The Potato Eaters*. The series depicts potato growers in France and are dark pieces of mostly browns and golds. The stark, irregular faces of the farmers reveal a depth of toil unfamiliar to most in First World urban environments. They are not lovely to look at but they are arresting, and they give pride of place to potatoes and the people who were traditionally overlooked in the stratification of society. Van Gogh elaborated on his intent in his letters.

"I have wanted to stress the fact that these people, eating their potatoes by the light of their lamp, have themselves dug the earth with the very hands they are now putting into the dish; it thus speaks of work with the hands and of honestly earning the food that they eat."[87]

Years ago, when I took that darkroom photography course, I asked myself why the potato hadn't been included in more still life paintings. (Except for Van Gogh and Renoir, very few serious artists have painted the spud.) I realize now it's because that most art is looking for something that's pretty on the outside – form, colour, and line rather than

---

[87] Jan Hulsker, *The new complete Van Gogh : paintings, drawings, sketches* (Amsterdam: J.M Meulenhoff, 1996) 172.

symbol. What Van Gogh saw in the potato eaters was that beauty exists not just in surface appearance. It exists in relationships, too – interpersonally as well as with the environment.

"I wanted it to give the idea of a way of life quite different from ours, of civilized people," wrote Van Gogh. "What I absolutely would not want is for everyone to consider it just beautiful or good."[88]

When Fred asked Desi how he got into the potato business, he replied, "Well, just like everyone else," he said. But, just how an enterprising young man went about using spuds to get through school is not at all obvious to people today, and neither is how communities can take back their food. If factory farms and GM potatoes existed back then, it's unlikely these memories would have been created. But if they could've been, the quality of the exchange and transmission of culture would surely be tempered by the relationship that the farmer has with his "art." No longer would s/he be invested in it in the same way. When you have to take directions from a third party on how to grow potatoes, or when you have to pour chemicals on a seed potato to make it grow, this shift in the process is bound to have an effect on how those farmers perceive themselves, and how we create the stories that we draw nourishment from. Artistry, mastery, and community are things we are losing as our food production becomes more industrialized and genetically modified. The art of living is damaged in the process, too. That's why it's so important to listen to what the potato has to say.

The spud's role in our lives is not just connected to culture and identity (to whatever degree), it's not just about tasting or looking good, it's about how it can help us feed ourselves without damaging anyone or thing in the process. A business that requires you to take food out of people's

[88] Hulsker, 172.

mouths or make plants sterile just can't be a dignified way to make your living. If such a person's life were a painting, what would it look like, I wonder? I'm sure it wouldn't be beautiful even in form, line or colour.

Raymond Loo wondered, too, about the choices society seems to be making.

"I think it's presumptuous of us. This mad cow business, [for example]. It shows you [what happens] if you take animals and put them in a non-natural environment and feed them unnatural feed. Basically everything is completely unnatural compared to what they'd be when the hunters and gatherers were there, (not that I want to be back there), but we do it the way it's completely convenient for us to do. But we still expect the food to be healthy for us! And I think the same thing for potatoes and all the different vegetables. If you take them and you basically completely change the natural environment [by destroying the biodiversity] and just keep the potato, [that means] you're feeding them chemically. [So,] why would we expect this food to be good for us? You know, it's kind of a leap. I mean, if you looked at any other thing besides food, people would say, well, I'm not gonna do that. You know, it wouldn't make sense, but for some reason this whole industrialization of the food industry, it's like a train that keeps on going."

Stopping the Train

It was obvious from my visit to the PoPa that the Quechua have put up their hand and said, *The train stops here!* But the conception and management of the PoPa is much more organic than are Western economic practices in that everything arises from their culture's laws and by consensus amongst the communities that thrive within the park. It was also clear that there was a built-in mechanism for making sure that the good of the community was served and if not, measures were taken to address that. It wouldn't be a bad model to base the laws on in North America: if corporations are doing things that do not help the community as a whole, then they must be taken to task. Taking that idea even further, a system that made sure such things wouldn't happen at all would be even better. When Schmeiser won his case he didn't say the ruling would ensure what happened to him would never be repeated. He said "when" it happened, meaning that it's obvious GM contamination will continue unchecked unless we take serious action to check it. Lawsuits won't stop it. People will – when they put in place a system that safeguards and nurtures the community instead of taking advantage of it. What a novel idea.

Individualism

While it's clear how the Quechua have gone about trying to change things, it's not clear how this approach could be adapted to the Western world where individualism is both a blessing and a curse.

"The European culture is all about individualism," said Raymond Loo, "and of course Canada and the United States were based on that. And we're not good team players. As farmers we're not good team players. In the Federation, they're always fighting, instead of getting a bunch of people together, and I'm not sure exactly why. I think as long as our success is measured in the bank account instead of being measured in environmental stewardship, for example, that's going to continue. Everybody's trying to get an edge on other people. But I saw it when I was in Guatemala, too. They have organic cooperatives that work together down there. As individuals they couldn't afford the money, the wherewithal, the land base, and so on, to be individually certified, but as a group, it works. When you try to get them together in North America, it doesn't work.

For some reason or other, we don't have the mentality, right now."

Stephen Bell noted the same difficulty in Northern Ireland. "I think if anything could ever put more money into potatoes it would be sharing equipment," he said, "and trying to work together in some sort of better cooperation. I think in NI there are more combine harvesters than there are fields. Everyone who grows cereals will have a combine. And potato growers are probably similar. They all have their own equipment rather than sharing.

"Why don't they share their farm equipment?" asked Fred.

"They're so independent, and they wouldn't like anyone else knowing what they're doing. This is mine, and it's really like a status symbol ... but, you'll never know if they're not successful," unless they go bankrupt.

John Shepherd owns all his own equipment, "because the day you need it done, somebody else will need it done. We haven't enough good weather here to spend three to four weeks with the potatoes. It's got to be now; it's a rush match ... To me, the only way forward is stay small, and do it all yourself."

The mentality of individualism is tied to historian Salaman's point regarding class disparity where the potato is concerned, except that the power structure has shifted in our time to that of corporations instead of specific nations. Today the entity that subjugates is more likely to be a large business rather than a country. Free trade, mono-crops, patents, and GMOs have shifted things and now it's people everywhere (including the likes of Raymond Loo, John Shepherd and us, not just people from developing countries) who are put in the position of the weaker class.

Doing it all yourself may be the Western way, but it won't work if our food sources become irrevocably damaged by biotech companies seeking bigger profit margins for their shareholders. If biotech companies like Monsanto continue succeeding, it will be down to the bystanders who did nothing as well as the shareholders who also profit from the colonization of our food and our DNA. "In 2007, Monsanto employed 18,000 workers in 50 countries. In 2007, its stock prices continue to rise and its profits have reached a billion dollars. Its shareholders include not only pension funds and banks, but also hundreds of thousands of small investors."[89]

Tools to rewrite this script are available in the Western world, but they all require ordinary people getting involved. As an anonymous blogger puts it, "Those who don't take an active interest in politics will forever be ruled by those that do."[90] While we have been going about our daily lives, the movers and the shakers of the political and business worlds have made shocking changes to the landscape we all depend upon and in which we all have the right to thrive. Besides getting involved in social justice groups, like those listed in the appendix, spud lovers can get involved with potato breeding clubs. Plant scientist Raoul Robinson, who is convinced that individuals can indeed make a huge difference, has written *The Amateur Potato Breeders' Manual* in which he makes a complicated topic very easy to understand.

"The people most likely to be interested in potato breeding clubs," writes Raoul, "are the organic farmers, as well as anyone who is worried about the high cost of potatoes, and their contamination with both pesticides and genetic engineering. Anyone who thinks that amateur breeders cannot achieve these objectives should consider

---

[89] *The World According to Monsanto*

[90] Anonymous blogger, ‹www.livevideo.com/Smokescreen?subtype=1›.

the success that amateurs have had in breeding marijuana. Motivation is the prime incentive.[91]

Other options are community organic gardens and food banks. One such group in Toronto, The Stop Community Food Centre, is in the process of building "The Green Barn," a huge facility that will include an indoor and outdoor garden, soil creation area, and community kitchen. "It will be an incredible place to bring people together to show them how to grow food," said Nick Saul, the centre's executive director. Organizations like these underscore the importance of community. Growing food goes hand in hand with growing a culture that supports and nourishes people and nature in respectful, reciprocal ways.

However, as important as getting involved at the grassroots' level is, if biotech companies are allowed to continue disseminating their GM seed, no amount of legal recourse will put back into Pandora's Box all the ills released into our food system. People have to make their voices heard strongly enough for governments to listen and change. We did it when we in North American let it be known that we did not want GM potatoes in our fast food. McCain, along with many other fast food chains, listened. But the train has not yet slowed, never mind stopped! In fact, UK scientists recently said that they want to keep the public in the dark regarding GM field trials they are conducting, "just like Canada does." At a press conference in July 2008, "scientists said the number of field trials had declined in recent years because of sabotage, damaging the UK's ability to inspire innovation and commercial investment."[92]However, as Drs. Pusztai and Ewen pointed

---

[91] Raoul A. Robinson, *The Amateur Potato Breeder's Manual.* ‹www.sharebooks.ca/eBooks/SpudsManual.pdf›.

[92] "Researchers want crop trial sites kept secret," *Daily Mail*, 28 July 2008 ‹www.dailymail.co.uk/sciencetech/article-1039332/Researchers-want-crop-trial-sites-kept-secret-beat-anti-GM-protestors.html#›.

out, commercial interests are not ones that can be relied upon to give full and accurate information about food and health safety. That scientists should have to rely on businesses rather than impartial governments to fund their experiments says that governments are not doing their job of looking after our best interests, which is what we elect them to do in the first place.

More and more governments are working for big business rather than their citizens, and more and more, the public is being kept in the dark. That scientists supposedly working for the public good now want to add to that situation is alarming.

In Britain, "protesters are able to find the sites in the UK because their location is publicly available under rules brought in to allow farmers to know what was being grown near them – but in Canada small-scale trials which are judged not to have environmentally damaging consequences are not publicized."[93] Also not publicized until after the fact was the 1999 secret fast-tracking between Health Canada and the Canadian Food Inspection Agency (CFIA), and Monsanto of two of that biotech company's genetically modified potato varieties. "When the CFIA discovered the "extremely poor" field tests they asked Monsanto for more data – but Monsanto refused. Health Canada stated, 'Monsanto objected to these requests, believing that their data adequately supports their conclusions that these products present 'no significant environmental, feed or food safety risk.' To obtain the data from Monsanto, the departments struck a deal with the corporation where they

---

[93] "Researchers want crop trial sites kept secret," Daily Mail, 28 July 2008 ‹www.dailymail.co.uk/sciencetech/article-1039332/Researchers-want-crop-trial-sites-kept-secret-beat-anti-GM-protestors.html#›.

pledged to speedily decide on the approval of the potato – within 30 days of receiving the information."[94]

With dealings like this going on behind our backs, it's obvious that we are still being used as guinea pigs in the GM experiment where the whole world is its laboratory.

Sexy, eh?

There is a lot to do and a lot to learn from each other. The stories of Alejandro Argumedo, Raymond Loo, and Arpad Pusztai are particularly inspiring and light the way for others.

When Raymond explained to me that each true seed of a potato could grow into any number of potato varieties, I was delighted at the marvel that was the humble spud. I was as pleased as if I had discovered buried treasure in my grandfather's backyard. I did. It was the potato.

Thanks to Alejandro's subtle acuity, I will never look at a potato the same way again. Stopping to show us the Sacred Valley has reprogrammed my Western worldview. Now when I see piles of potatoes, I think of Mother Earth's spud-brown breasts resting on the shores of the river, water lapping against her life-giving nipples.

When I visited Arpad and Susan, they fed me with organic food they either grew themselves or bought from local farmers they knew well. In fact, we ate some prosciutto-style meat that came from a pig they knew by name. We drank homemade, organic wine that Susan's uncle Sandor had made. Afterwards, Arpad took me for a walk down the hill to Lake Balaton. He told me of the *fogas* in the lake, the perch fish found only in that body of water. This reedy lake, like Lake Titicaca, is steeped in history. It has seen the Romans come and go. It has seen relatives from the East and West of Germany reunite before the Wall fell. It has seen the Magyars create a fertile wine region out of what once was desert. The Hungarian word for

---

[94] Lucy Sharratt, "One Potato."

desert is "puszta," intriguingly similar to Arpad's last name, Pusztai. Interestingly, the name "Arpad" belonged to a famous Magyar King. Given how courageous Arpad has been in the wake of his career's destruction, it was no surprise to me that he would have sprung from a land that had been transformed for the common good.

This book has taken me on a journey that helped me hear what the spud was singing to me at my grandmother's table. As I ate the potatoes she grew and prepared with her own hands, I realize now that my taste buds sang with the exuberantly rich potential inside every person. We may not all be born of kings, but we can do royally good things, and that's no small potatoes!

Appendix – Just a Few Organizations …

Guelph Organics Conference www.guelphorganicconf.ca	*This conference is for the industry as well as the public. Seminars, talks, magazines, and over 150 exhibitors (like Geneaction and CBAN) make it a great place to get information and meet like-minded people.*
Canadian Biotechnology Action Network 431 Gilmour Street, Second Floor, Ottawa, Ontario, K2P 0R5 info@cban.ca	*Collaborative campaigning for food sovereignty and environmental justice*
Canadian Health Coalition www.healthcoalition.ca	*A not-for-profit, non-partisan organization dedicated to protecting and expanding Canada's public health system.*
Canadian Organic Growers National Office 323 Chapel Street Ottawa, ON K1N 7Z2 office@cog.ca	*Canadian Organic Growers Inc. is Canada's national membership-based education and networking organization representing farmers, gardeners, consumers and supporters.*
Arpad Pusztai Page www.plab.ku.dk/tcbh/Pusztaitcbh.htm	*A comprehensive site about Dr. Pusztai's GM potato study*
ETC Group Headquarters 431 Gilmour St, 2nd Floor Ottawa, ON K2P 0R5 etc@etcgroup.org	*ETC monitors power, tracks technology, and supports diversity.*

Raymond Loo info@springwillowfarms.com	*Breeder of Island Sunshine potatoes and other organic produce.*
IFOAM Head Office Charles-de-Gaulle-Str. 5 53113 Bonn Germany headoffice@ ifoam.org	*An umbrella group with hundreds of member associations worldwide, it promotes organic agriculture.*
Irish Organic Farmers and Growers Association Main Street Newtownforbes Co. Longford, Ireland info@iofga.org	*Certifying the organic integrity of foodstuffs, produce, and products for farmers, growers, food processors, wholesalers, traders, and retailers. They also have a magazine, Organic Matters.*
Rodale Institute 611 Siegfriedale Road Kutztown, PA 19530-9320 USA Phone: 610-683-1400 www.rodaleinstitute.org	*Rodale has 60 years of sustainable farming experience and information available to farmers and the public.*
The Centre for Food Safety 600 Pennsylvania Ave. SE #302 Washington, DC 20003 office@centerforfoodsafety.org	*A non-profit public interest and environmental advocacy membership organization*

Selected Works Cited

Donnelly, James S. Jr. *The Great Irish Potato Famine.* Gloucestershire: Sutton Publishing, 2004.

Salaman, Redcliffe N. *The history and social influence of the potato.* Cambridge: University Press, 1949.

Shiva, Vandana. *Biopiracy: The Plunder of Nature and Knowledge.* Boston: South End Press, 1997.

—. *Monocultures of the Mind: Perspectives on Biodiversity and Biotechnology.* London: Zed Books, 2000.

—. *Stolen Harvest: The Hijacking of the Global Food Supply.* Cambridge: South End Press, 2000.

Smith, Jeffrey M. *Genetic Roulette: The Documented Health Risks of Genetically Engineered Foods.* Fairfield: Yes! Books, 2007.

The World According to Monsanto. Dir. Marie-Monique Robin. 2008.

Index

I

J

L

M

N

P

Q

R

S

T

U

W

Printed in the United States
123315LV00001B